# Reactive Patterns with RxJS and Angular Signals

Elevate your Angular 18 applications with RxJS Observables, subjects, operators, and Angular Signals

**Lamis Chebbi**

# Reactive Patterns with RxJS and Angular Signals

**Group Product Manager**: Kaustubh Manglurkar

**Publishing Product Manager**: Vaideeshwari Muralikrishnan

**Senior Editor**: Hayden Edwards

**Technical Editor**: Simran Ali

**Copy Editor**: Safis Editing

**Project Coordinator**: Shagun Saini

**Proofreader**: Safis Editing

**Indexer**: Rekha Nair

**Production Designer**: Jyoti Kadam

**Marketing Coordinator**: Anamika Singh and Nivedita Pandey

First edition published: April 2022

Second edition published: July 2024

Production reference: 1290524

Published by Packt Publishing Ltd

Grosvenor House

11 St Paul's Square

Birmingham

B3 1RB

ISBN 978-1-83508-770-1

www.packtpub.com

*To my father, who instilled in me diligence, perseverance, and a good work ethic. Thank you for always being there to support me and lift me up.*

*To my mother, who taught me selflessness and doing things with love. Thank you for your enduring encouragement during the writing of this book.*

*To my brother and my sisters, for their continuous support.*

*– Lamis Chebbi*

# Foreword

RxJS is a powerful JavaScript library that enables developers to build reactive and event-based web applications. The Angular framework uses this library to manage asynchronous operations, such as HTTP communication and user interaction with web forms and routing.

Angular Signals, a cutting-edge API, introduces fine-grained reactivity in Angular applications. This synchronous reactive pattern boosts performance and intelligently tracks the application state, optimizing component rendering and enhancing the overall user experience.

*Reactive Patterns with RxJS and Angular Signals* is a book that embraces both worlds, combining best practices from each tool to help you build performant and reactive Angular applications.

Lamis uses a simple yet insightful approach to RxJS and Signals, one which not only allows you to gain a deeper understanding of all the available reactive patterns in Angular, but also one that helps you build a complete application that encompasses all the latest features of the Angular framework.

*Aristeidis Bampakos*

*Angular Google Developer Expert (GDE)*

# Contributors

## About the author

**Lamis Chebbi** is a Google Developer Expert for Angular and is the author of the first edition of this book, titled *Reactive Patterns with RxJS for Angular*. She is an enthusiastic software engineer with a strong passion for the modern web, the founder of Angular Tunisia, a member of the WWCode community, a speaker, a content creator, and a trainer.

She has been interested in Angular and RxJS for the past few years and loves to share her knowledge about Angular by participating in workshops and organizing training sessions. Empowering women and students is one of her highest priorities.

Besides Angular and the web, Lamis loves music, traveling, chromotherapy, and volunteering.

Last but not least, she's a forever student.

*I want to thank all the people who believed in me, supported me, and inspired me throughout this journey.*

# About the reviewers

**Aleksandr Guzenko** is a respected software engineer known for his extensive expertise in both frontend and backend development. With over eight years of experience in the field, Aleksandr has made significant contributions to the software engineering community. His deep knowledge and practical skills are not only evident in his professional work but also in his active involvement as a judge in numerous hackathons.

Furthermore, Aleksandr is a sought-after speaker at conferences, where he shares his insights and experiences, particularly focusing on software architecture. He is also an author, of articles that delve into various aspects of software engineering.

**Ishu Mishra** is an IT specialist, working on frontend tech stacks for over 6 years. He began working at a start-up organization called Sparx IT Solutions, where he gained knowledge of Angular, and has worked for several companies since then.

Ishu previously worked for the Japanese multinational NEC Corporation, and presently, Ishu is employed in Bangalore by Morgan Stanley.

**Matheus Rian** is a frontend developer and speaker, who is passionate about technology and education. He started programming in high school and hasn't stopped learning since, developing his skills in technologies such as Angular and React.

Furthermore, Matheus is a multiplier, seeking to disseminate knowledge and generate a positive impact within communities. As well as working in the frontend market, he gives lectures at events and contributes articles on Medium and `dev.to`.

**Arthur Lannelucq** is a passionate frontend developer specializing in Angular and RxJS. He is a strong advocate for reactive architectures, believing in their power to build responsive and scalable web applications.

He has gained solid experience working on various projects for large companies and start-ups. Eager to share his knowledge, he also runs a YouTube channel where he provides tutorials and practical advice on Angular and frontend development.

# Table of Contents

# Part 2: A Trip into Reactive Patterns

3

## Fetching Data as Streams    25

4

## Handling Errors Reactively    53

# 5

## Combining Streams                                                  71

# 6

## Transforming Streams                                               87

# 7

## Sharing Data between Angular Components                            111

# Part 3: The Power of Angular Signals

8

# Part 4: Multicasting Adventures

9

## 10

## 11

## 12

# Part 5: Final Touches

## 13

## Testing RxJS Observables    207

# Preface

Embarking on the journey from imperative to reactive programming is a significant shift and one that I have experienced firsthand. As I navigated this transition, I found myself drawn to the world of reactive patterns and the transformative power they held. It was a journey filled with discovery, comparison, and a strong determination to understand this new way of thinking.

Inspired by my own experiences, I've crafted this book to serve as a guide through the realms of reactive patterns within Angular applications. I believe that the reactive mindset is gradually achieved by comparing the reactive way to the imperative way, in order to distinguish the difference and benefits. Within these pages, you'll discover how embracing reactive patterns can greatly enhance the way you manage data, write code, and react to user changes. From improving efficiency to creating cleaner, more manageable code bases, the benefits are vast and practical.

So, without further ado, let's embark on this journey together and unlock the potential of reactive programming.

## Who this book is for

If you're a developer working with Angular and RxJS, this book is tailor-made for you. Designed for individuals at a beginner level in both Angular and RxJS, this book will guide you toward becoming an experienced developer while also benefitting those who wish to harness the potential of RxJS and leverage the reactive paradigm within their Angular applications.

## What this book covers

In *Chapter 1*, *Diving into the Reactive Paradigm*, you will learn the fundamentals of reactive programming.

In *Chapter 2*, *Walking through Our Application*, you will learn the architecture and requirements of the recipe application that we will be building through the book.

In *Chapter 3*, *Fetching Data as Streams*, you will learn the reactive pattern for fetching data so that we can reactively retrieve a list of recipes in our recipe app.

In *Chapter 4*, *Handling Errors Reactively*, you will learn the different error strategies and the reactive patterns for handling errors.

In *Chapter 5*, *Combining Streams*, you will learn the reactive pattern for combining streams and use it to implement a filter functionality in our recipe app, while also discovering the common pitfalls and sharing best practices for optimal implementation.

In *Chapter 6, Transforming Streams*, you will learn the reactive pattern for transforming streams and use it to implement autosave and autocomplete features in our recipe app.

In *Chapter 7, Sharing Data between Angular Components*, you will learn the reactive pattern to share data between components and use it to share the selected recipe in our recipe app.

In *Chapter 8, Mastering Reactivity with Angular Signals*, you will deep-dive into Angular signals, learning different reactive patterns based on Angular Signals, and how to unleash the power of RxJS and Signals together. You will also discover the latest Angular Signals improvements.

In *Chapter 9, Demystifying Multicasting*, you will learn the essentials of multicasting and the different multicasting concepts and operators offered by RxJS, such as Subjects, Behavior Subjects, and Replay Subjects.

In *Chapter 10, Boosting Performance with Reactive Caching*, you will learn the reactive pattern to cache streams and implement a caching mechanism in our recipe app, based on the latest RxJS features.

In *Chapter 11, Performing Bulk Operations*, you will learn the reactive pattern to perform bulk operations and implement a multiple asynchronous file upload in our recipe app.

In *Chapter 12, Processing Real-Time Updates*, you will explore the reactive patterns to consume real-time updates and display newly created recipes instantly in our recipe app.

In *Chapter 13, Testing RxJS Observables*, you will learn the different strategies to test reactive patterns and practice testing the API responses in our recipe app.

## To get the most out of this book

This book assumes some familiarity with Angular, basic RxJS, TypeScript, and a foundational knowledge of functional programming. All code examples have been tested using Angular 17 and 18 on the Windows OS. However, they should work with future version releases too.

| Software/hardware covered in the book | Operating system requirements |
| --- | --- |
| Angular 17 and above | Windows, macOS, or Linux |
| TypeScript 5.4.2 | Windows, macOS, or Linux |
| RxJS 7.8.1 | Windows, macOS, or Linux |
| PrimeNG 17.10.0 | Windows, macOS, or Linux |
| Bootstrap 5.0.0 | Windows, macOS, or Linux |

Make sure you follow the prerequisites found here: `https://angular.dev/tools/cli/setup-local`. The prerequisites include the environment setup and the technologies needed in order to install and use Angular.

We also use the Bootstrap library to manage the application's responsiveness, the PrimeNG library for its rich components, and, of course, RxJS as the reactive library.

Plus, there is a ready-for-use backend server in the GitHub repository that we will only reference in our application.

**If you are using the digital version of this book, we advise you to type the code yourself or access the code from the book's GitHub repository (a link is available in the next section). Doing so will help you avoid any potential errors related to the copying and pasting of code.**

## Download the example code files

You can download the example code files for this book from GitHub at `https://github.com/PacktPublishing/Reactive-Patterns-with-RxJS-and-Angular-Signals-Second-Edition`. If there's an update to the code, it will be updated in the GitHub repository.

We also have other code bundles from our rich catalog of books and videos available at `https://github.com/PacktPublishing/`. Check them out!

## Conventions used

There are a number of text conventions used throughout this book.

`Code in text`: Indicates code words in text, database table names, folder names, filenames, file extensions, pathnames, dummy URLs, user input, and Twitter handles. Here is an example: "In the following code snippet, we have an example of an Angular service that injects the `HttpClient` service and fetches data from the server using the `HttpClient.get()` method."

A block of code is set as follows:

```
import { Injectable } from '@angular/core';
import { HttpClient } from '@angular/common/http';
import { Observable} from 'rxjs';
```

Any command-line input or output is written as follows:

```
//console output
Full Name: John Doe
```

**Bold**: Indicates a new term, an important word, or words that you see on screen. For instance, words in menus or dialog boxes appear in **bold**. Here is an example: "Users can create a new recipe by clicking on the **New Recipe** menu item located at the top right of the page."

> **Tips or important notes**
> Appear like this.

# Get in touch

Feedback from our readers is always welcome.

**General feedback**: If you have questions about any aspect of this book, email us at customercare@ packtpub.com and mention the book title in the subject of your message.

**Errata**: Although we have taken every care to ensure the accuracy of our content, mistakes do happen. If you have found a mistake in this book, we would be grateful if you would report this to us. Please visit www.packtpub.com/support/errata and fill in the form.

**Piracy**: If you come across any illegal copies of our works in any form on the internet, we would be grateful if you would provide us with the location address or website name. Please contact us at copyright@packtpub.com with a link to the material.

**If you are interested in becoming an author**: If there is a topic that you have expertise in and you are interested in either writing or contributing to a book, please visit authors.packtpub.com.

# Share Your Thoughts

Once you've read *Reactive Patterns with RxJS and Angular Signals*, we'd love to hear your thoughts! Scan the QR code below to go straight to the Amazon review page for this book and share your feedback.

https://packt.link/r/1-835-08770-1

Your review is important to us and the tech community and will help us make sure we're delivering excellent quality content.

# Download a free PDF copy of this book

Thanks for purchasing this book!

Do you like to read on the go but are unable to carry your print books everywhere?

Is your eBook purchase not compatible with the device of your choice?

Don't worry, now with every Packt book you get a DRM-free PDF version of that book at no cost.

Read anywhere, any place, on any device. Search, copy, and paste code from your favorite technical books directly into your application.

The perks don't stop there, you can get exclusive access to discounts, newsletters, and great free content in your inbox daily

Follow these simple steps to get the benefits:

1.  Scan the QR code or visit the link below

https://download.packt.com/free-ebook/9781835087701

2.  Submit your proof of purchase
3.  That's it! We'll send your free PDF and other benefits to your email directly

# Part 1:
# An Introduction
# to the Reactive World

Embark on a journey into the world of reactive programming with Angular!

In this part, you will understand the fundamentals of the reactive paradigm and its application in Angular, gaining insight into why it's essential to leverage this approach. Then, we will introduce the recipe application that we are going to progressively build as we go through the book.

This part includes the following chapters:

- *Chapter 1, Diving into the Reactive Paradigm*
- *Chapter 2, Walking through Our Application*

# 1

# Diving into the Reactive Paradigm

Reactive patterns are reusable solutions to a commonly occurring problem using reactive programming. Behind all these patterns is a new way of thinking, a new architecture, new coding styles, and new tools. That's what this entire book is based on – useful reactive patterns in Angular applications.

Now, I know you are impatient to write your first reactive pattern in Angular, but before doing so, and in order to help you take full advantage of all the RxJS patterns and leverage the reactive paradigm, we will start by explaining in detail all the fundamentals and preparing the groundwork for the following chapters.

Let's start with a basic understanding of the reactive paradigm, its advantages, and the problems it solves. Best of all, let's put a reactive mindset on and start thinking reactively. We will begin by highlighting the pillars and the advantages of the reactive paradigm. Then, we will explain the marble diagram and why it is useful. Finally, we will highlight the use of RxJS in Angular.

Giving an insight into the fundamentals of the reactive paradigm is incredibly important. This will ensure you get the basics right, help you understand the usefulness of the reactive approach, and consequently help you determine which situation is best to use it in.

In this chapter, we're going to cover the following topics:

- Exploring the pillars of reactive programming
- Learning the marble diagram (our secret weapon)
- Highlighting the use of RxJS in Angular

## Technical requirements

This chapter does not require any environment setup or installation steps.

All the code snippets in this chapter are just examples to illustrate the concept, so you will not need the code repository to follow along. However, if you're interested, the code for the book can be found at `https://github.com/PacktPublishing/Reactive-Patterns-with-RxJS-and-Angular-Signals-Second-Edition`.

This book assumes that you have a basic understanding of Angular and RxJS.

> **Note**
>
> This book uses the new Angular documentation site, `angular.dev`. The previous documentation site, `angular.io`, will soon be deprecated. Stay connected with the latest updates and resources by accessing the documentation through this link.

## Exploring the pillars of reactive programming

**Reactive programming** is among the major programming paradigms used by developers worldwide. Every programming paradigm solves some problems and has its own advantages. By definition, reactive programming is programming with asynchronous data streams and is based on observer patterns. So, let's talk about these pillars of reactive programming!

### Data streams

**Data streams** are the spine of reactive programming. Everything that may change or happen over time (you don't know when exactly) is represented as asynchronous streams such as events, notifications, and messages. Reactive programming is about reacting to changes as soon as they are emitted!

An excellent example of data streams is UI events. Let's suppose that we have an HTML button and we want to execute an action whenever a user clicks on it. Here, we can think of the click event as a stream:

```
//HTML code
<button id='save'>Save</button>

//JS code
const saveElement = document.getElementById('save');
saveElement.addEventListener('click', processClick);

function processClick(event) {
  console.log('Hi');
}
```

As implemented in the preceding code snippet, in order to react to the click event, we register an `EventListener` event. Then, every time a click occurs, the `processClick` method is called to execute a side effect. In our case, we are just logging `Hi` in the console.

As you might have gathered, to be able to react when something happens and execute a side effect, you should listen to the streams to become notified. To get closer to reactive terminology, instead of *listen*, we can say *observe*. This leads us to the *observer* design pattern, which is at the heart of reactive programming.

## Observer patterns

The **observer pattern** is based on two main roles – a publisher and a subscriber:

- A **publisher** maintains a list of subscribers and notifies them or propagates a change every time there is an update
- On the other hand, a **subscriber** performs an update or executes a side effect every time they receive a notification from the publisher

The observer pattern is illustrated here:

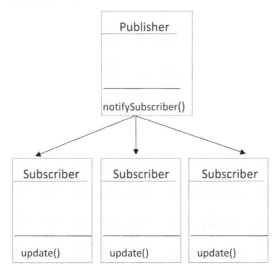

Figure 1.1 – The observer pattern

To get notified about the updates, you need to subscribe to the publisher. A real-world analogy would be a newsletter; you don't get any emails from a specific newsletter if you don't subscribe to it.

This leads us to the building blocks of RxJS, which include the following:

- **Observables**: These are a representation of the asynchronous data streams that notify the observers of any change
- **Observers**: These are consumers of the data streams emitted by Observables

RxJS combines the observer pattern with the iterator pattern and functional programming to process and handle asynchronous events. This was a reminder of reactive programming fundamentals, and it is crucial to know when to put a reactive implementation in place and when to avoid it.

In general, whenever you have to handle asynchronous tasks in your Angular application, always think of RxJS. The main advantages of RxJS over other asynchronous APIs are as follows:

- RxJS makes dealing with event-based programs, asynchronous data calls, and callbacks an easy task.
- Observables guarantee consistency. They emit multiple values over time so that you can consume continuous data streams.
- Observables are lazy; they are not executed until you subscribe to them. This helps with writing declarative code that is clean, efficient, and easy to understand and maintain.
- Observables can be canceled, completed, and retrieved at any moment. This makes a lot of sense in many real-world scenarios.
- RxJS provides many operators with a functional style to manipulate collections and optimize side effects.
- Observables push errors to the subscribers and provide a clean way to handle errors.
- RxJS allows you to write clean and efficient code to handle asynchronous data in your application.

Now that we have given some insight into the reactive programming pillars and detailed the major advantages of RxJS, let's explore the marble diagram, which is very handy for understanding and visualizing the Observable execution.

## Learning about the marble diagram (our secret weapon)

RxJS ships with more than one hundred **operators** – these are among the building blocks of RxJS, useful for manipulating streams. All the reactive patterns that will be detailed later in this book are based on operators, and when it comes to explaining operators, it is better to refer to a visual representation – that's where marble diagrams come in!

**Marble diagrams** are visual representations of the operator's execution, which will be used in all chapters to understand the behavior of RxJS operators. At first, it might seem daunting, but it is delightfully simple. You only have to understand the anatomy of the diagram and then you'll be good at reading and translating it.

Marble diagrams represent the execution of an operator, so every diagram will include the following:

- **Input Observable(s)**: Represents one or many Observables given as input to the operator
- **Operator**: Represents the operator to be executed with its parameters
- **Output Observable**: Represents the Observable produced after the operator's execution

We can see the execution illustrated here:

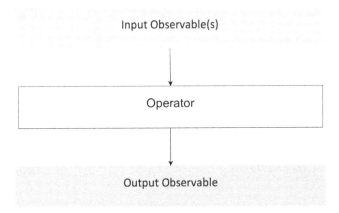

Figure 1.2 – The operator execution

Now, let's zoom in on the representation of the input/output Observables:

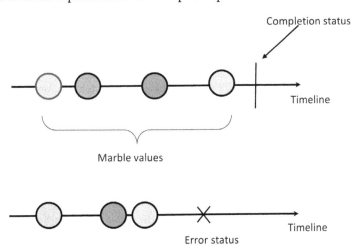

Figure 1.3 – The marble diagram elements

The elements of these diagrams include the following:

- **The timeline**: Observables are asynchronous streams that produce data over time. Therefore, the representation of time is crucial in the marble diagram, and it is represented as an arrow flowing from left to right.

- **The marble values**: These are the values emitted by the Observables over time. They are represented by colored circles.

- **The completion status**: The vertical line (|) represents the successful completion of the Observables.

- **The error status**: The **X** represents an error emitted by the Observable. Neither the values nor the vertical line representing completion will be emitted thereafter.

That's all the elements you need to know about. Now, let's put all the pieces together in a real marble diagram:

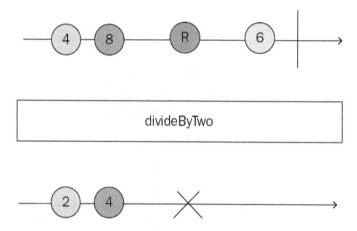

Figure 1.4 – An example of a marble diagram for a custom operator

As you may have guessed, we have a custom operator called `divideByTwo` that will emit half of every received number. When the input Observable emits the values 4 and 8, the output Observable produces 2 and 4 respectively.

However, when the R value, which is non-numeric, is emitted, then an error is thrown, indicating abnormal termination. This case is not handled in the operator code. The input Observable continues the emission and then completes successfully. However, the value will never be processed because, after the error, the stream is closed.

At this point, we've gone through all the elements composing the marble diagram. You will be able to understand the operators used in the chapters to come. Now, let's shed some light on the use of RxJS in Angular.

# Highlighting the use of RxJS in Angular

RxJS is practically a first-class citizen in Angular. It is part of the Angular ecosystem and is used in many features to handle asynchronous tasks. The following are some examples of these features:

- The HttpClient module
- The Router module
- Reactive forms
- The Event emitter

We will discuss each of the following concepts in the subsequent subsections.

> **Note**
>
> We recommend taking a quick look at `https://angular.dev/overview`, where you can find further details about the aforementioned features.

## The HttpClient module

You might be familiar with the HttpClient API provided by Angular that is used to communicate with a server over the HTTP protocol. The `HttpClient` service is based on Observables that manage all transactions, which means that the result of calling API methods such as GET, PATCH, POST, and PUT will be an Observable.

In the following code snippet, we have an example of an Angular service that injects the `HttpClient` service and fetches data from the server using the `HttpClient.get()` method:

```
import { Injectable } from '@angular/core';
import { HttpClient } from '@angular/common/http';
import { Observable} from 'rxjs';
import { Recipe } from '../model/recipe.model';
@Injectable()
export class RecipesService {
constructor(private http: HttpClient) { }
getRecipes(): Observable<Recipe[]> {
return this.http.get<Recipe[]>(`api/recipes/`);
}
}
```

The `getRecipes()` method – or, to be more accurate, the call to `this.http.get<Recipe>(`api/recipes/`)` – returns an Observable that you should subscribe to in order to send the GET request to the server. Please note that this is an example of an HTTP transaction, and it is the same for all the other HTTP methods available in the API (POST, PUT, PATCH, etc.).

> **Note**
>
> The code contains references to `recipe.model` and `getRecipes()` – in *Chapter 2, Walking through Our Application*, you will be introduced to the Recipe app that we will be working on throughout the rest of the book.

For those familiar with Promise-based HTTP APIs, you may be wondering about the advantages of using Observables in this context. For those who are not acquainted with Promises, **Promises** are JavaScript objects that represent the eventual completion (or failure) of an asynchronous operation and its resulting value. They provide a cleaner and more structured way to work with asynchronous code compared to traditional callback-based approaches. However, there are a lot of advantages of using Observables over Promises, and the most important ones are listed as follows:

- Observables are cancellable, so you can cancel the HTTP request whenever you want by calling the unsubscribe method
- You can also retry HTTP requests when an error occurs or an exception is thrown

The server's response cannot be mutated by Observables, although this can be the case for chaining `then()` on Promises.

## The Router module

The Router module available in the `@angular/router` package uses Observables in router events and activated routes. We will look at both here.

### *Router events*

Router events allow you to intercept the navigation life cycle. They are defined in the Router as Observables.

> **Note**
>
> We recommend taking a quick look at `https://angular.dev/api/router/Event`, where you can find further details about router events.

The majority of Angular applications have a routing mechanism. Router events change frequently over time, and it makes sense to listen to changes to execute side effects. That's why Observables are a flexible way in which to handle those streams.

To intercept all the events the router goes through, first, you should inject the `Router` service, which provides URL manipulation capabilities. Then, subscribe to the events Observable available in the `Router` object and filter the events of type `RouterEvent` using the RxJS filter operator.

The following is an example of an Angular service that injects the Router in the constructor, subscribes to the router events, and traces the event ID and path in the console:

```
import { Injectable } from '@angular/core';
import { Router, RouterEvent } from '@angular/router';
import { filter } from 'rxjs/operators';
@Injectable()
export class CustomRouteService {
  constructor(public router: Router) {
    this.router.events.pipe(
      filter(event => event instanceof RouterEvent)
    ).subscribe((event: RouterEvent) => {
      console.log(`The current event is : ${event.id} |
        event.url`);
    });
  }
}
```

This is a very basic example, and you could introduce pretty much any specific behavior to it.

### The activated route

ActivatedRoute is a router service that you can inject into your components to retrieve information about a route's path and parameters. Many properties are based on Observables. Here, you will find the implementation of the ActivatedRoute class:

```
class ActivatedRoute {
  snapshot: ActivatedRouteSnapshot
  url: Observable<UrlSegment[]>
  params: Observable<Params>
  queryParams: Observable<Params>
  fragment: Observable<string | null>
  data: Observable<Data>
  outlet: string
  component: Type<any> | string | null
  routeConfig: Route | null
  root: ActivatedRoute
  parent: ActivatedRoute | null
  firstChild: ActivatedRoute | null
  children: ActivatedRoute[]
  pathFromRoot: ActivatedRoute[]
  paramMap: Observable<ParamMap>
```

```
    queryParamMap: Observable<ParamMap>
    toString(): string
}
```

As you may have figured out, url, params, queryParams, fragment, data, paramMap, and queryParamMap are represented as Observables. All these parameters might change over time, so it makes sense to listen to these changes to register side effects or update the values.

Here's an example of an Angular component that injects the ActivatedRoute class in the constructor and then in ngOnInit() method, subscribes to the following properties:

- The url property of ActivatedRoute, in order to log the current URL in the console
- The queryParams property of ActivatedRoute, in order to retrieve the criteria parameter and store it in a local property named criteria:

```
import { Component, OnInit } from '@angular/core';
import { ActivatedRoute } from '@angular/router';

@Component({
  selector: 'app-recipes',
  templateUrl: './recipes.component.html'
})
export class RecipesComponent implements OnInit {
  criteria: string;
  constructor(private activatedRoute: ActivatedRoute) { }

  ngOnInit() {
    this.activatedRoute.url
      .subscribe(url => console.log('The URL changed to: '
        + url));

    this.activatedRoute.queryParams.subscribe(params => {
      this.processCriteria(params.criteria);
    });
  }
  processCriteria(criteria: string) {
    this.criteria = criteria;
  }
}
```

This example showcases the usage of the url and queryParams properties. For a comprehensive overview of all ActivatedRoute properties and their functionalities, I encourage you to visit the Angular documentation page at https://angular.dev/api/router/ActivatedRoute#properties.

## Reactive forms

Reactive forms available under the @angular/forms package are based on Observables to track form control changes. Here's the overview of the FormControl class in Angular:

```
class FormControl extends AbstractControl {
//other properties here
valueChanges: Observable<any>
statusChanges: Observable<any>
}
```

The properties valueChanges and statusChanges of FormControl are represented as Observables that trigger change events. Subscribing to a FormControl value change is a way of triggering application logic within the component class.

Here's an example that subscribes to the valueChanges Observable of a FormControl property called rating and simply traces the value through console.log(value):

```
import { Component, OnInit } from '@angular/core';
import { FormGroup } from '@angular/forms';
@Component({ ...})
export class MyComponent implements OnInit {
  form!: FormGroup;

  ngOnInit() {
    const ratingControl = this.form.get('rating');
    ratingControl?.valueChanges.subscribe(
      (value) => {
        console.log(value);
      }
    );
  }
}
```

This way, you will get the changed value as an output.

## The Event emitter

The event emitter, which is part of the @angular/core package, is used to emit data from a child component to a parent component through the @Output() decorator. The EventEmitter class extends the RxJS subject and registers handlers for events emitted by this instance:

```
class EventEmitter<T> extends Subject {
  constructor(isAsync?: boolean): EventEmitter<T>
  emit(value?: T): void
```

```
subscribe(next?: (value: T) => void, error?: (error: any)
  => void, complete?: () => void): Subscription
}
```

This is what happens under the hood when you create an event emitter and emit a value.

The following code block is an example of emitting the updated value of a recipe rating:

```
import { Component, Output } from '@angular/core';
import { EventEmitter } from 'events';

@Component({ ...})
export class RecipesComponent {
  constructor() {}
  @Output() updateRating = new EventEmitter();

  updateRecipe(value: string) {
    this.updateRating.emit(value);
  }
}
```

So, the `EventEmitter` smooths communication between components by allowing one component to emit custom events and another component to listen for and respond to those events. This mechanism plays a crucial role in enabling parent-child communication, sibling communication, and even communication between unrelated components in Angular applications.

> **Note**
>
> In the previous code snippets, the subscription to the Observables was done explicitly for demonstration purposes. In a real-world example, we should include the unsubscription logic if we want to subscribe explicitly. We will shed light on this in *Chapter 3, Fetching Data as Streams*.

## Summary

In this chapter, we walked you through the fundamentals of reactive programming and in which use cases it shines. Then, we explained the marble diagram that will be our reference for explaining RxJS operators in all the following chapters. Finally, we highlighted the use of reactive programming in Angular by illustrating concrete examples, implementations, and advantages.

Now that we have got the basics right, it is time to start preparing and explaining, in the next chapter, the application that we are going to build throughout this book, where we are going to implement all of the reactive patterns we will learn progressively.

2

# Walking through
# Our Application

Now, we are one step closer to diving into reactive patterns, but before we do, let's present the app that we are going to build throughout this book.

We will start by explaining the technical requirements, followed by a breakdown of the app's interfaces so that you know its user story. Furthermore, we will showcase an overview of the application architecture and a visual representation of the component tree. By the end of this chapter, we will have all the required pieces in place to start implementing our application.

In this chapter, we're going to cover the following main topics:

- Breaking down our app's interfaces
- Reviewing our app's architecture
- Reviewing our app's components

## Technical requirements

Though we are not creating the project in this chapter, you should know the requirements for it before going ahead.

We are going to use **Angular 18** for our frontend, so please make sure you follow the prerequisites at `https://angular.dev/tools/cli/setup-local`. The prerequisites include the environment setup and the technologies needed in order to install and use Angular.

We are also going to be using **Bootstrap** version 5.0.0 (`https://getbootstrap.com/`), a toolkit for developing responsive web apps, and version 7.8.1 of **RxJS**.

You will also be able to find all of the code to create this project in the book's GitHub repository: `https://github.com/PacktPublishing/Reactive-Patterns-with-RxJS-and-Angular-Signals-Second-Edition`.

# Breaking down our app's interfaces

As a food junkie, I want the application to be like a recipe book, allowing users and home cooks to browse and share delicious food recipes. The main aim of the app is to provide inspiration for meals as well as help users do the following:

- Share their recipes
- Pin favorite recipes to easily find them
- Distinguish top-rated recipes
- Filter out recipes according to some criteria

The app is composed of six interfaces. Let's tackle these interfaces one by one.

## View one – the landing page

The first page contains a list of available recipes, sorted according to popularity:

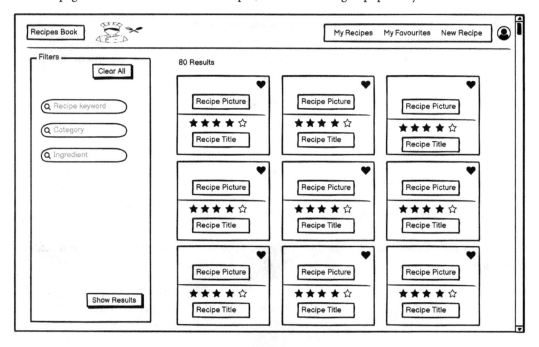

Figure 2.1 – The landing page view

In this view, users have the possibility to do the following:

- Quickly search for a recipe by setting filters according to some criteria (on the left-hand side)
- Clear the list of filters
- View the most popular recipes
- Rate recipes by clicking on the number of stars
- Add a recipe to their favorites by clicking on the heart icon
- See a recipe's details by clicking on it
- View the total number of recipes

## View two – the New Recipe interface

This page contains a form to create a new recipe:

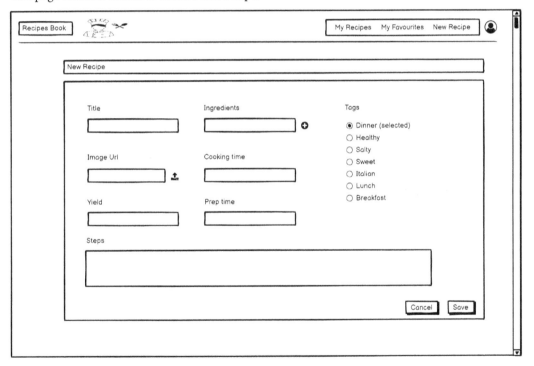

Figure 2.2 – The New Recipe view

In this view, users can create a new recipe by clicking on the **New Recipe** menu item located at the top right of the page. A form containing details of the recipe will be opened to fill out the information and save it. These details include the following:

- **Title**: The title of the recipe
- **Ingredients**: The ingredients required to prepare the recipe
- **Image Url**: A good image of the meal prepared
- **Cooking time**: The time required to cook the meal
- **Yield**: The number of people that can be served by this meal
- **Prep time**: The time required to prepare the meal
- **Tags**: Key tags describing the recipe
- **Steps**: The steps required to prepare and cook the meal

## View three – the My Recipes interface

This page contains a list of recipes created by the user. This screen is reachable by clicking on the **My Recipes** menu item located at the top right. Users can edit and remove recipes by clicking on the edit and delete icons, respectively:

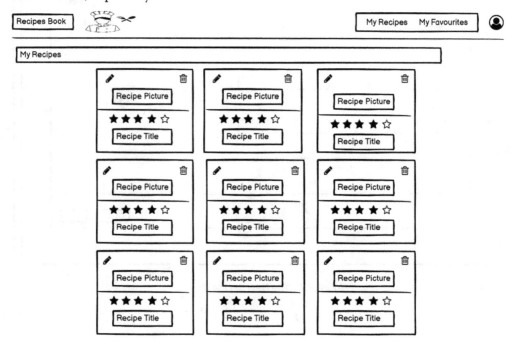

Figure 2.3 – The My Recipes view

## View four – the My Favourites interface

This page contains a list of the user's favorite recipes, reachable by clicking on the **My Favourites** menu item located at the top right:

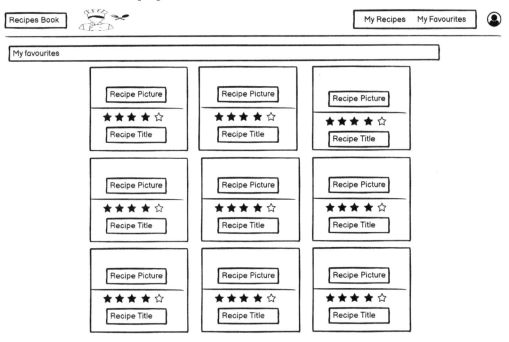

Figure 2.4 – The My Favourites view

## View five – the Modify Recipe interface

While the **New Recipe** interface allows the user to create a new recipe, the **Modify Recipe** interface allows the user to edit an existing recipe. This page is reachable by clicking the **Edit** button next to each button on the **My Recipes** interface, and looks just like *Figure 2.2*.

### View six – the Recipe Details interface

This page contains all the details of the selected recipe. This screen is reachable from the landing page after clicking on a displayed recipe:

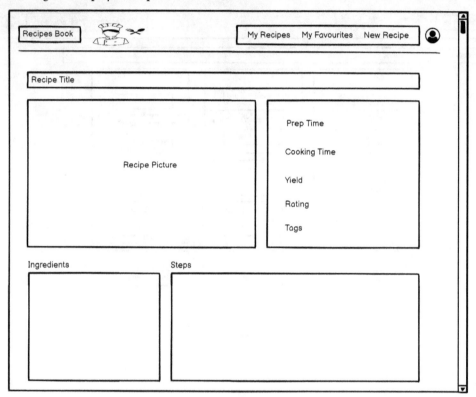

Figure 2.5 – The Recipe Details view

Now that we have detailed our application's interfaces, let's have a look at the app's architecture.

## Reviewing our app's architecture

The frontend layer of the recipe app will be implemented in Angular 18 and will communicate with a RESTful backend based on Node.js.

> **Note**
>
> Aspects related to the backend are not the subject of this book and will not be detailed. You can find a ready-to-use fake backend, named `recipes-book-api`, in the GitHub repository: `https://github.com/PacktPublishing/Reactive-Patterns-with-RxJS-for-Angular-17-2nd-Edition`.

The frontend of the recipe app is pluggable to any RESTful backend. Therefore, you can use pretty much any other technology for the backend. All communications will be performed through the HttpClient module and will request REST controllers in the backend:

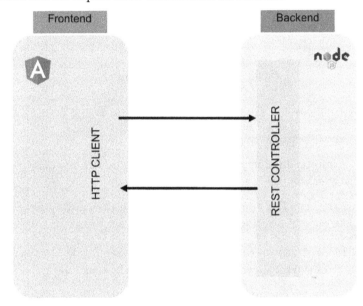

Figure 2.6 – The book of recipes architecture

Now that we have the big picture of our target application, let's break down the different Angular components of our app.

## Reviewing our app's components

An Angular application has a tree structure consisting of all the components we create. In the following diagram, you will find the component tree of our recipe app, which is important for understanding the anatomy of the application:

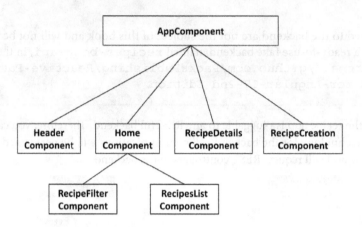

Figure 2.7 – Components overview

Let's break down the components:

- AppComponent: The parent component of the app

  HeaderComponent: The component representing the header of the app that contains the user space, the menu, and the logo

- HomeComponent: The component representing the landing page that contains RecipeFilterComponent and RecipesListComponent:

  - RecipeFilterComponent: The component representing the filter zone that contains the criteria fields and the **Clear All** and **Show Results** buttons

  - RecipesListComponent: The component containing a list of recipes

- RecipesDetailsComponent: The component containing the details of one recipe

- RecipesCreationComponent: The component containing a form to create a recipe with all the required fields

You now have a better understanding of the components that will make up our app.

## Summary

In this chapter, we explained the features of the recipe application that we will be working on, as well as the look and feel of the UIs. We also shed light on the app's architecture and the components that will make it up.

Now that all those aspects are clear, let's explore our first reactive pattern together, which we will cover in the next chapter.

# Part 2:
# A Trip into
# Reactive Patterns

In this part, you will learn the most used reactive patterns in different real-world scenarios such as fetching data from a backend API, handling server errors, filtering data, and providing autocompleted search results in a dropdown list. Every reactive pattern will be endorsed by an example involving our recipes app.

You will also learn the best practices and pitfalls to avoid and dive into the latest Angular features, such as standalone components and the new built-in control flow syntax.

This part includes the following chapters:

- *Chapter 3, Fetching Data as Streams*
- *Chapter 4, Handling Errors Reactively*
- *Chapter 5, Combining Streams*
- *Chapter 6, Transforming Streams*
- *Chapter 7, Sharing Data between Angular Components*

# 3

# Fetching Data as Streams

The way you manage your application's data has a huge impact on your UI performance and the user experience. As far as I'm concerned, great user experience and performant UIs are no longer an option nowadays – they are key determinants of user satisfaction. Furthermore, managing data efficiently optimizes the code and enhances its quality, which consequently minimizes maintenance and improvement costs.

So, how can we manage our data efficiently? Well, this is what we will be answering in the following chapters. There are a few reactive patterns that come in handy in many use cases, and we will start by exploring the most basic reactive pattern for displaying values received from a REST endpoint to allow users to read and interact with them.

To begin, we will explain the requirement that we're going to implement in the recipe application. Then, we will introduce the classic pattern to retrieve data, followed by the different approaches that you can use to manage unsubscriptions and all the important technical concepts surrounding them. We will also learn about a new feature of Angular 14+, which is standalone components. Following this, we will explain the reactive pattern to fetch data, and highlight the advantages of the reactive pattern over the classic one. Finally, we will learn about the new built-in control flow introduced in Angular 17.

So, in this chapter, we're going to cover the following main topics:

- Defining the data fetch requirement
- Exploring the classic pattern for fetching data
- Exploring the reactive pattern for fetching data
- Highlighting the advantages of the reactive pattern
- Diving into the built-in control flow in Angular 17

# Technical requirements

This chapter assumes that you have a basic understanding of HttpClient, Angular components, Angular modules, and routing.

We'll be using a mocked REST API backend built with JSON Server, which allows you to spin up a REST API server with a fully working API. We'll not be learning how to use JSON Server, but if you are interested in learning more, you can find further information at `https://github.com/typicode/json-server`.

You can access the project source code for this chapter in the GitHub repository at `https://github.com/PacktPublishing/Reactive-Patterns-with-RxJS-and-Angular-Signals-Second-Edition/tree/main/Chap03`.

The project is composed of two folders:

- `recipes-book-api`: This contains the mocked RESTful server already set up.
- `recipes-book-front`: This contains the frontend application that was built with Angular 17 and RxJS 7. As a third-party dependency, we've added `bootstrap` and `primeng` libraries to help us build beautiful UI components quickly. Please refer to the previous chapter's *Technical requirements* section for the environment and dependencies setup.

The project complies with the standard Angular style guide, which can be found at `https://angular.dev/style-guide#`.

The first time you run the apps, you will need to install dependencies beforehand. You only have to run the `npm i` command in the `recipes-book-api` and `recipes-book-front` folders.

Once dependencies are installed, you need to start the server by running the following command in the `recipes-book-api` folder:

```
npm run server: start
```

The server will be running at `http://localhost:8081`.

Then, you start the frontend by running the following command in the `recipes-book-front` folder:

```
ng serve --proxy-config proxy.config.json
```

You can read more about the `--proxy-config` parameter at `https://angular.dev/tools/cli/serve#proxying-to-a-backend-server`.

# Defining the data fetch requirement

First, let's define the requirement we are going to implement in a reactive way. We want to display the list of recipes retrieved from the mocked backend on the home page, progressively building the user story detailed in the *View one – the landing page* section of *Chapter 2, Walking through Our Application*:

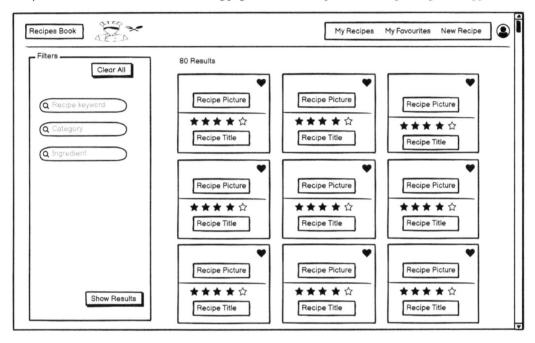

Figure 3.1 – The landing page view

To do this, we need to fetch the list of recipes beforehand to display it to the user as cards, right? So, the list of recipes represents the first data that we need to request, which is available in our recipes-book-api server through the following endpoint:

```
GET /api/recipes
```

Please don't forget to start the server as detailed in the *Technical requirements* section. Once the server is started, you can check the result of the fetch API at http://localhost:8081/api/recipes.

In the following sections, we will see how we can implement the fetching data requirement in both classic and reactive styles to understand the basic differences between them, and to see the benefits the reactive programming gives us over the imperative one.

# Exploring the classic pattern for fetching data

Let's first have a look at the implementation of the classic pattern for fetching the list of recipes.

## Defining the structure of your data

First and foremost, we need to define the structure of our data so that we can strongly type it. This will allow us to take advantage of TypeScript's type-checking features and catch type errors early.

We can use the Angular CLI to generate the **Recipe** model underneath the `src/app/ core/ model` folder:

```
$ ng g i Recipe
```

For convention purposes, we will change the name of the generated file from `recipe.ts` to `recipe. model.ts`. Then, we will populate the interface with the specific properties of `Recipe`, as follows:

```
export interface Recipe {
id: number;
title: string;
ingredients: string;
tags?: string;
imageUrl: string;
cookingTime?: number;
prepTime?: number;
yield: number;
steps?: string;
rating:number;
}
```

One by one, we enter the properties of the recipe we are going to use, followed by the type of each property. The description of each property is detailed in *View two – the New Recipe interface* section of *Chapter 2, Walking through Our Application*.

For optional properties, we placed a question mark (?) just before the property's type annotation when declaring the interface to tell TypeScript that the property is optional.

## Creating the fetching data service

The next step is to create an Angular service named `RecipesService` that will be responsible for managing all the operations around the recipes. This service will encapsulate the **create, read, update, and delete (CRUD)** operations and make them available to the various UI components. In this chapter, we will only implement the read (fetch) operation.

Now, why do we create a service? Well, we do it to increase modularity and to ensure the reusability of the service over the other components.

To generate the service underneath the `core/services` folder, we execute the `ng g s` command under the `core/services` folder like so:

```
$ ng g s Recipes
```

Now that the service is generated successfully, let's create and implement the method that will have the responsibility of fetching the data. We will inject the `HttpClient` in the `RecipesService` and define a method to retrieve the data. The service will look like this:

```
import { Injectable } from '@angular/core';
import { HttpClient } from '@angular/common/http';
import { Observable } from 'rxjs';
import { Recipe } from '../model/recipe.model';
import { environment } from 'src/environments/environment';
const BASE_PATH = environment.basePath

@Injectable({
providedIn: 'root'
})

export class RecipesService {

constructor(private http: HttpClient) { }

getRecipes(): Observable<Recipe[]> {
return this.http.get<Recipe[]>(`${BASE_PATH}/recipes`);
}
}
```

Let's break down what is going on at the level of `RecipesService`. It's nothing fancy – we have a `getRecipes()` method that gets the list of recipes over HTTP and returns a strongly typed HTTP response: `Observable<Recipe[]>`. This Observable represents the data stream that will be created when you issue the HTTP GET request. When you subscribe to it, it will emit the list of recipes as a JSON array and then completes. So, the stream represented by this HTTP request will be completed after emitting the response data once.

As a best practice, we externalized BASE_PATH in the `environment.ts` file because, in many cases, the server's base path depends on the environment (such as testing or production). This way, it is easier to update the paths in one place rather than updating all the services using it.

> **Note**
>
> Starting from Angular 15, the environment files are not shipped by default anymore. However, you can choose to generate them on demand by executing `$ ng g environments`.

We've also injected the `HttpClient` dependency in the constructor as follows:

```
constructor(private http: HttpClient) { }
```

This technique is known as **constructor injection**. Angular's built-in dependency injection system will automatically provide the injected dependencies when an instance of the component or service is created.

Plus, starting from version 14, Angular's dependency injection system provided the `inject()` utility function. This allows you to manually resolve and retrieve dependencies within a component or a service like so:

```
private http= inject(HttpClient);
```

This approach is useful when you need to dynamically resolve dependencies or perform conditional dependency injection based on runtime conditions.

We will use the constructor injection technique throughout the book. However, should you wish to adopt the newer approach, you have the flexibility to do so.

## Creating Angular standalone components

Now, we should create the component responsible for displaying the list of recipes named `RecipesListComponent` under `src/app/recipe-list`. Before that, though, let's just stop here and explain a super interesting new type of component introduced in Angular 14: the standalone component.

By definition, a **standalone component** is a self-contained component that doesn't belong to any `NgModule` and can be used by either other standalone components or module-based components.

Before Angular 14, we only had one way to create components:

```
$ ng g c recipesList
```

This command will create a component named `RecipesListComponent` and add it to an `NgModule`. Which module, though? If you specify `--module` in the command line, followed by the path of your module, then the CLI will add the component in that specific module. If the `--module` option is not set, then the CLI will check whether there's a module in the same directory; if not, it will check in the nearest parent directory. If neither of those options is the case, it will generate a new module file in the same directory as the component and declare the component in that new module.

In short, the CLI will always end up associating the component to a module and adding it to the module's declaration array; otherwise, you will get a compilation error.

However, starting from Angular 14, you can decide to create a standalone component that doesn't belong to any NgModule by mentioning the --standalone flag in the command line:

```
$ ng g c recipesList --standalone
```

Using this in our project, `RecipesListComponent` won't be added to an `NgModule`, and will contain the `standalone: true` flag inside the `@Component` decorator, as well as the `imports` property:

```
@Component({
  selector: 'app-recipes-list',
  standalone: true,
  imports: [],
  templateUrl: './recipes-list.component.html',
  styleUrls: ['./recipes-list.component.scss'],
})
```

If the standalone component depends on other components, whether module-based or standalone, you should mention those components in the `imports` array; otherwise, you will get a compilation error.

Standalone components can also be used by module-based components or other standalone components. Plus, they can be used when loading routes, and in lazy loading. It's also worth knowing that you can create standalone directives and standalone pipes as well.

So far, so good! Now, why should you care? There are a few good reasons we should adopt standalone components in our projects:

- Less code means less boilerplate to write, and hence, quicker build times, plus better code organization, testing, and maintainability.

- It is easier to understand the component's dependencies as they are mentioned directly in the `imports` property of the standalone component. For a module-based component, you will have to scan your component's code and then check the module's dependencies that are shared by all the components belonging to that module.

- The power of standalone components lies in their isolation and self-contained nature. You only import what is needed by your component, while module-based components sometimes import useless dependencies used by other components in the same module.

Let's suppose that we have a module "M" that imports "A," "B," and "C" components and "S1," "S2," and "S3" services, and we have a "D" component that does not belong to that module but depends on the component "B." As "B" is a module-based component, then "D" should import the entire module 'M'; this leads to unnecessary dependencies as "D" does not need the components "A" and "B" or the services "S1" and "S2." So, integrating standalone components gives us more flexibility to only import the components and services required, since standalone components are self-contained and have their own sets of dependencies and logic. Consequently, it eliminates redundant code, leading to a more optimized app.

- It makes the learning curve for beginner Angular developers less steep.

We will be using standalone components in our recipe app to adopt a modular and self-contained approach. We will only keep the app component as a module-based component, even though we can bootstrap the application using a standalone component. Here's a schema representing our component's dependencies:

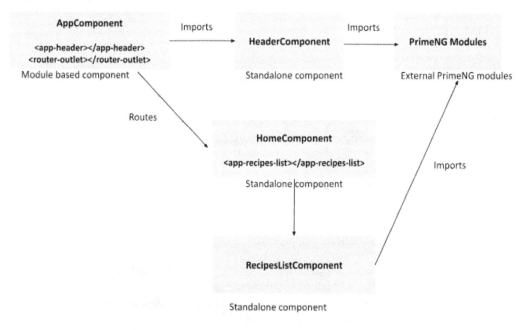

Figure 3.2 – The recipe app's components' dependencies

The `AppComponent` parent component is a module-based component that imports the `HeaderComponent` standalone component in the `AppModule` imports declaration. `HeaderComponent` uses some PrimeNG external dependencies, so it needs to be imported in the component's imports declaration.

HomeComponent is a standalone component that will be routed to by AppComponent. HomeComponent imports the RecipesListComponent standalone component in the component's imports declaration. The latter uses some PrimeNG external dependencies, so it needs to be imported in the component's imports declaration. All the code is available in the GitHub repository.

> **Note**
>
> For more information on standalone components, you can check https://angular.dev/ reference/migrations/standalone

Hopefully, the concept of standalone components is clear, so let's move on to the following step.

## Injecting and subscribing to the service in your component

In this section, we will inject the RecipesService service in the RecipesListComponent component and call the getRecipes() method in ngOnInit() (when the component is initialized). We will also make a read operation against the API server.

In order to get the data emitted, we need to subscribe to the returned Observable from the getRecipes() method. Then, we bind the data to a local array property created in our component, called recipes. The component's code will look like this:

```
import { Component, OnInit } from '@angular/core';
import { Observable } from 'rxjs';
import { Recipe } from '../core/model/recipe';
import { RecipesService } from '../core/services/recipes.
Service';


@Component({
selector: 'app-recipes-list',
standalone: true,
imports: [CommonModule],
templateUrl: './recipes-list.component.html',
styleUrls: ['./recipes-list.component.scss']
})
export class RecipesListComponent implements OnInit {

recipes!: Recipe[];
constructor(private service: RecipesService) { }

ngOnInit(): void {
this.service.getRecipes().subscribe(result => {
this.recipes = result;
```

```
  });
  }
  }
```

Now that we've retrieved the data and stored it in a local property, let's see how we will display it in the UI.

## Displaying the data in the template

Now we can use the `recipes` property (which is available in the component) in our HTML template to display the list of recipes in our UI. In our case, we are using the `DataView` PrimeNG component to display the list of recipes as cards in a grid layout (further details about this component can be found at `https://primeng.org/dataview`).

Of course, our goal is to focus not on the template code, but on the manipulation of the data inside it. As you can see in the following example, we passed the `recipes` array to the `value` input of the data view component (you can also use structural directives to render a data view component with pure HTML if you don't want to include a third-party dependency):

```
<div class="card">
<p-dataView #dv [value]="recipes" [paginator]="true"
[rows]="9"    filterBy="name" layout="grid">
/** Extra code here **/
</p-dataView>
</div>
```

This is the basic pattern for collecting data, which you would have discovered back when you started learning about Angular, so you have likely seen something like this before.

Now there's just one thing left – you should handle the unsubscription of the Observable, as this code manages subscriptions manually. Otherwise, the `Observable` subscription will stay alive after the component has been destroyed, and the memory's reference will not be released, causing memory leaks. That's why you should always be careful of this when manually subscribing to Observables inside Angular components.

> **Note**
>
> Although `HttpClient` Observables unsubscribe automatically after the server request responds or times out, we will still demonstrate how to handle their unsubscription to secure our implementation and showcase best practices. This will also serve as a showcase for handling unsubscription with other Observables.

## Managing unsubscriptions

There are two commonly used ways to manage unsubscriptions: the imperative pattern and the declarative reactive pattern. Let's look at both patterns in detail.

### *Imperative unsubscription management*

Imperative unsubscription means that we manually call the `unsubscribe()` method on the subscription object that we manage ourselves. The following code snippet illustrates this:

```
export class RecipesListComponent implements OnInit,
OnDestroy {
  recipes!: Recipe[];
  subscription: Subscription;
  constructor(private service: RecipesService) { }

ngOnInit(): void {
  this.subscription=this.service.getRecipes()
  .subscribe(result => {
    this.recipes = result;
});
}

ngOnDestroy(): void {
  this.subscription?.unsubscribe();
}
```

Here, we simply store the subscription inside a variable called `subscription` and unsubscribe from it in the `ngOnDestroy()` lifecycle hook.

This works fine, but it is not a recommended pattern. There is a better way, using the power of RxJS.

### *Declarative unsubscription management*

The second unsubscription method is cleaner and far more declarative, using the RxJS `takeUntil` operator. However, before we dive into this pattern, let's gain an understanding of the role of `takeUntil` using the following marble diagram:

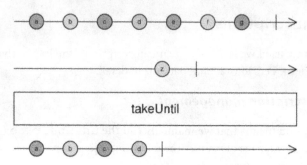

Figure 3.3 – The takeUntil marble diagram

The takeUntil() operator emits values from the source Observable (the first timeline) until the Observable notifier, which is given as input (the second timeline), emits a value. At that time, takeUntil() will stop the emission and complete. In the marble diagram, the source Observable emitted the values of a, b, c, and d – so takeUntil() will emit them, respectively. After that, the Observable notifier emits z, then takeUntil() will stop emitting values and will be completed.

In our application, the takeUntil operator will help us keep the subscription alive for a period that we define. We want it to be alive until the component has been destroyed, so we will create an RxJS subject that will emit a value when the component has been destroyed. Then, we will pass this subject to takeUntil as input:

```
export class RecipesListComponent implements OnInit,
OnDestroy {
  recipes!: Recipe[];
  destroy$ = new Subject<void>();
  constructor(private service: RecipesService) { }

ngOnInit(): void {
  this.service.getRecipes().pipe(
    takeUntil(this.destroy$)).
    subscribe(result => {
    this.recipes = result;
  });
}

ngOnDestroy(): void {
  this.destroy$.next();
  this.destroy$.complete();


}
}
```

> **Note**
>
> The $ sign is an informal convention that is used to indicate that the variable is an Observable.

The first thing you might notice here is that it's less code than the first approach. Furthermore, when we call `unsubscribe()` on a returned subscription object (the first way), there's no way we can be notified that the unsubscription happened. However, using `takeUntil()`, we will be notified of the Observable completion through the completion handler.

It is worth noting that this implementation can be further enhanced by using the `takeUntilDestroyed` operator introduced in Angular 16. This operator simplifies Observable subscription management in your Angular components and directives. It automatically completes subscriptions when the associated component or directive is destroyed, eliminating the need for manual cleanup in the `ngOnDestroy` lifecycle hook.

You only have to import the `takeUntilDestroyed` operator from the `@angular/core/rxjs-interop` package as follows:

```
import { takeUntilDestroyed } from '@angular/core/rxjs-interop';
```

Then, we use this operator within the pipe operator of our subscription. The previous code will look like this after using `takeUntilDestroyed`:

```
export class RecipesListComponent {
  recipes!: Recipe[];
  constructor(private service: RecipesService) {
    this.service.getRecipes().pipe(takeUntilDestroyed())
      .subscribe(result=>this.recipes = result);
  }
}
```

As you can see, the manual cleanup code in the `ngOnDestroy` lifecycle hook has been removed along with the `destroy$` subject, resulting in a more concise and readable component implementation.

The `takeUntilDestroyed()` operator will automatically handle the subscription cleanup when `RecipesListComponent` is destroyed.

Apart from the `takeUntil` and `takeUntilDestroyed` operators, there are other operators that manage unsubscription for you in a more reactive way. The following are some examples:

- `take(X)`: This emits $x$ values and then completes (will no longer emit values). For example, `take(3)` will emit three values from the given Observable and then complete. However, bear in mind that if your network is slow and the *xth* emission didn't happen, then you have to unsubscribe manually.

- `first()`: This emits the first value and then completes.

- `last()`: This emits the last value and then completes.

This was the classic pattern that we have all learned as a beginner, and it is a relatively valid way for fetching data. To sum up, the following diagram describes all the steps that we walked through:

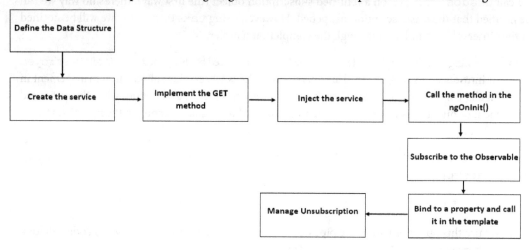

Figure 3.4 – The classic pattern workflow

However, there is another pattern that we can use, which is much more declarative and reactive and has many advantages. We'll discover it next!

# Exploring the reactive pattern for fetching data

The idea behind this reactive pattern is to keep and use the Observable as a stream throughout the application. Don't worry – this will become more apparent to you as you explore this section. Let's get started.

## Retrieving data as streams

To start using the reactive pattern, instead of defining a method to retrieve our data, we will declare a variable inside our service:

```
import { Injectable } from '@angular/core';
import { HttpClient } from '@angular/common/http';

import { Recipe } from '../model/recipe';
import { environment } from 'src/environments/environment';
const BASE_PATH = environment.basePath
```

```
@Injectable({
providedIn: 'root'
})
export class RecipesService {

recipes$ = this.http.get<Recipe[]>(
`${BASE_PATH}/recipes`);
constructor(private http: HttpClient) { }
}
```

Here, we are declaring the `recipes$` variable as the result of HTTP GET, which is either an Observable or the data stream. Think of every piece of data that changes over time as a stream and declare it as an Observable in a separate service. This will make it accessible throughout the app and give us more flexibility to manipulate it in different parts of the application.

## Defining the stream in your component

Now, in `RecipesListComponent`, we are going to do the same thing we did in the classic pattern – that is, declare a variable holding the stream returned from our service. However, this time, the variable is the Observable we created in `RecipesService`:

```
import { Component, OnDestroy, OnInit } from '@angular/core';
import { RecipesService } from '../core/services/recipes.
Service';

@Component({
selector: 'app-recipes-list',
standalone: true,
imports: [CommonModule],
templateUrl: './recipes-list.component.html',
styleUrls: ['./recipes-list.component.css']
})
export class RecipesListComponent implements OnInit {

recipes$= this.service.recipes$;

constructor(private service: RecipesService) { }
}
```

But wait! We need to subscribe in order to get the emitted data, right? That's absolutely correct. Let's see how we will do it.

## Using the async pipe in your template

For this pattern, we will not subscribe manually but instead, use a better way, the async pipe. The **async pipe** makes rendering values emitted from the Observable easier.

First of all, it automatically subscribes to the input Observable. Then, it returns the latest value emitted. Best of all, when the component has been destroyed, it automatically unsubscribes to avoid any potential memory leaks. This means there is no need to manually clean up any subscriptions when the component has been destroyed. That's amazing!

So, in the template, we bind to an Observable using the async pipe. As `recipes` describes the array variable that the values are emitted into, we can use it in the template as follows:

```
<div *ngIf="recipes$ |async as recipes" class="card">
<p-dataView #dv [value]="recipes" [paginator]="true"
[rows]="9"    filterBy="name" layout="grid">
/** Extra code here **/
</p-dataView>
</div>
```

As you may have noticed, the `<div>` element contains a `*ngIf` structural directive. This directive conditionally renders its child elements based on the truthiness of the `recipes$ | async` expression.

The `recipes$ | async` expression subscribes to the `recipes$` Observable and asynchronously renders the child elements of the `<div>` element (which is the `DataView` component in our case) when the Observable emits a value. It also unsubscribes and cleans up the subscription when the element is removed from the **DOM** (**Document Object Model**).

The `*ngIf` directive is followed by `as recipes`, which assigns the emitted value from the Observable to the local `recipes` variable. This allows us to access the emitted value within the scope of the `<div>` element and its children using the `recipes` variable.

By using the async pipes, we don't need the `ngOnInit` lifecycle hook, as we will not subscribe to the `Observable` notifier in `ngOnInit()` and unsubscribe from `ngOnDestroy()` as we did in the classic pattern. Instead, we simply set a local property in our component and we are good to go – we don't need to handle the subscription and unsubscription on our own!

> **Note**
> The full code of the HTML template is available in the GitHub repository.

To sum up this pattern, the following diagram describes all the steps we walked through:

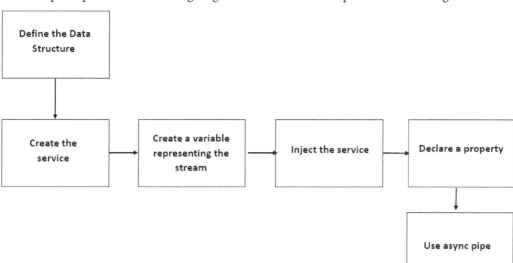

Figure 3.5 – The reactive pattern workflow

Now that we have explained the reactive pattern in action, in the next section, let's review its advantages.

# Highlighting the advantages of the reactive pattern

I think you might have guessed the first advantage of the reactive pattern – we don't have to manually manage subscriptions and unsubscriptions, and what a relief – but there are a lot of other advantages. Let's look at the other advantages in more detail.

## Using the declarative approach

Let's shed light on why we don't explicitly use the `subscribe()` method. What's wrong with `subscribe()`? Well, subscribing to a stream inside our component means we are allowing imperative code to leak into our functional and reactive code. Using the RxJS Observables does not make our code reactive and declarative systematically.

But what does declarative mean, exactly? Well, first, we will nail down some key terms. Then, let's iterate from there:

- A **pure function** is a function that will always return identical outputs for identical inputs, no matter how many times it is called. In other words, the function will always predictably produce the same output.

- **Declarative** refers to the use of declared functions to perform actions. You rely upon pure functions that can define an event flow. With RxJS, you can see this in the form of Observables and operators.

So, why should you care? Well, you should care because the declarative approach using RxJS operators and Observables has many advantages, namely, the following:

- It makes your code cleaner and more readable.

- It makes your code easier to test because it is predictable.

- It makes you able to cache the stream output given a certain input, and this will enhance performance. We will explore this in more detail in *Chapter 7, Sharing Data between Angular Components, Chapter 9, Demystifying Multicasting*, and *Chapter 10, Boosting Performance with Reactive Caching*.

- It enables you to leverage RxJS operators and transform and combine streams coming from different services or even within the same service. This is what we will see in *Chapter 5, Combining Streams*, and *Chapter 6, Transforming Streams*.

- It helps you react easily to user interactions in order to execute an action.

So, more declarative means more reactive. However, be careful. This doesn't mean you can't ever call the `subscribe()` method. It is unavoidable in some situations to trigger the `Observable` notifier. But try to ask yourself: do I really need to subscribe here? Can I instead compose multiple streams together, or use RxJS operators, to achieve what I need without subscribing? Aside from cases where it is unavoidable, never use `subscribe()`.

Now, let's move to the change detection concept and see how it can improve performance.

## Using the change detection strategy of OnPush

The other really cool thing is that we can use the `changeDetection` strategy, `OnPush`.

**Change detection** is one of the most powerful features of Angular. It is about detecting when the component's data changes and then automatically re-rendering the view or updating the DOM to reflect that change. The default strategy of "check always" means that, whenever any data is mutated or changed, Angular will run the change detector to update the DOM. So, it is automatic until explicitly deactivated.

In the OnPush strategy, Angular will only run the change detector when one of the following occurs:

- *Condition 1*: A reference of a component's @Input property changes (bear in mind that when the input property object is mutated directly, then the reference of the object will not change and, consequently, the change detector will not run. In this case, we should return a new reference of the property object to trigger the change detection).

- *Condition 2*: A component event handler is emitted or gets triggered.

- *Condition 3*: A bound Observable via the async pipe emits a new value.

Therefore, using the ChangeDetection OnPush strategy minimizes any change detection cycles and will only check for changes to re-render our components in the preceding cases. This strategy applies to all child directives and cannot be overridden.

In our scenario, we only need the change detector to run if we get a new value; otherwise, we get useless updates. So, our scenario matches *Condition 3*. The good news is that we can use the change detection onPush strategy as follows:

```
import { ChangeDetectionStrategy, Component} from
'@angular/ core';

@Component({
  selector: 'app-recipes-list',
  standalone: true,
  imports: [CommonModule],
  templateUrl: './recipes-list.component.html',
  styleUrls: ['./recipes-list.component.scss'],
  changeDetection: ChangeDetectionStrategy.OnPush
})
```

If we remember to use the async pipe as much as possible, we will see a couple of advantages:

- We will make it easier to later switch from the default change detection strategy to OnPush if we need to

- We will run into fewer change detection cycles using OnPush

In general, using the async pipe will help you to achieve a high-performing UI, and it will have a lot of impact if your view is doing multiple tasks.

And here's the output of our UI after all of that work in the chapter:

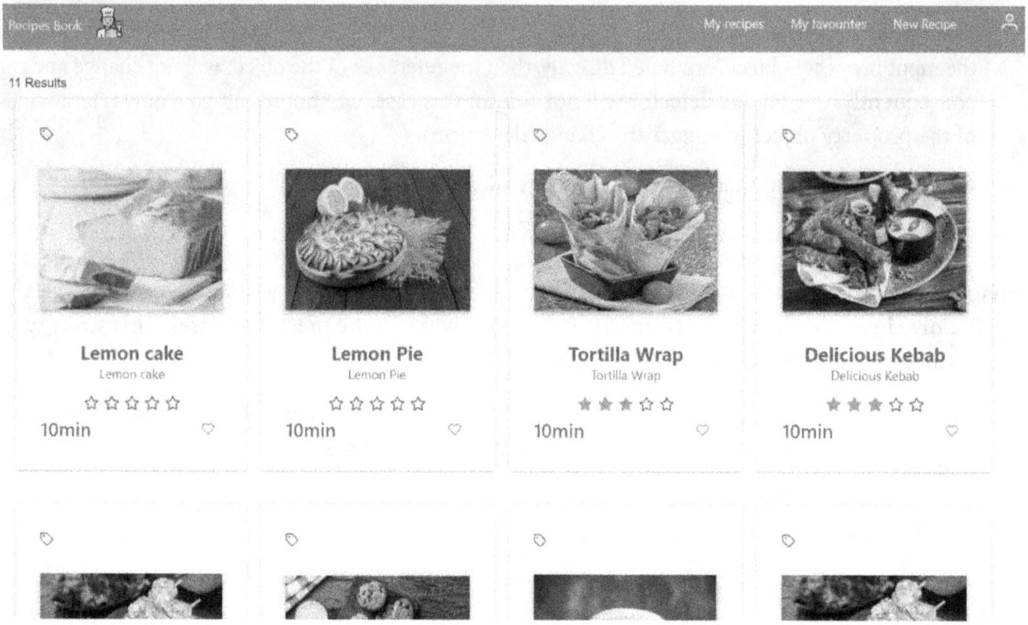

Figure 3.6 – An overview of the list of recipes

So, after all that, in a nutshell, using the reactive pattern for fetching data will improve the performance of your application, the change detection strategy, and the code clarity and readability. As well as that, it will make the code more declarative and reactive, it will make it easier to leverage RxJS operators, and it will make it easier to react to user actions.

Now that we've established the reactive pattern, let's conclude this chapter by exploring an intriguing feature introduced in Angular 17, understanding its benefits, and applying it in practice within our recipe app.

## Diving into the built-in control flow in Angular 17

Before Angular 17, control flow within templates was predominantly managed using structural directives. Let's start by exploring the structural directives.

## Structural directives

Structural directives are responsible for altering the structure of the DOM and orchestrating how elements are added, removed, or repeated based on certain conditions. Here's the list of available directives in Angular to control the execution of the template:

- *ngIf: This structural directive is used to conditionally include or exclude elements from the DOM based on the truthiness of an expression. For instance, consider the following code snippet, which displays the message **No items found** if the items array is empty:

```
<div *ngIf="items.length === 0">No items found </div>
```

To display alternative content when the condition is false, we can use an else statement, like so:

```
<div *ngIf="items.length === 0; else itemsFound">
  <div>No items found</div>
</div>
<ng-template #itemsFound>
  <div>Items found</div>
</ng-template>
```

In this code, if the items array is not empty, the content inside the else block defined by the ng-template element with the #itemsFound reference will be displayed, indicating **Items found**.

- *ngFor: This structural directive is used for iteration. It repeats a section of HTML for each item in an iterable collection. For example, this code renders a list of products one by one:

```
<ul>
  <li *ngFor="let product of products">
    {{ product.name }}
  </li>
</ul>
```

In order to improve performance, you can optionally add a custom trackBy function that provides a unique identifier for each item in the list. This is achieved by modifying the previous code, shown as follows:

```
<ul>
  <li *ngFor="let product of products; trackBy:
    trackProduct">{{ product.name }}
  </li>
</ul>
```

Then, define the `trackProduct` function in your component class to return the unique identifier of each product item as follows:

```
trackProduct(index: number, product: Product) {
  return product ? product.id : undefined;
}
```

This way, Angular can more efficiently track changes within the list. It will only update the DOM elements that actually changed, instead of re-rendering the entire list for minor changes. This leads to a smoother user experience, especially when dealing with large or frequently updated lists.

- `ngSwitch`: This structural directive is used to conditionally include or exclude elements from the DOM based on the evaluated value of a provided expression. It is commonly used when there are multiple conditions to be evaluated. Here's an example that renders different views based on user roles:

```
<div [ngSwitch]="userRole">
  <admin-dashboard *ngSwitchCase="admin" >
    </admin-dashboard>
  <user-dashboard *ngSwitchCase="'user'" >
    </user-dashboard>
  <guest-dashboard *ngSwitchDefault >
    </guest-dashboard>
</div>
```

Now that we've explored the structural directives in Angular, which provided a mechanism for dynamically altering the structure of the DOM based on certain conditions, we can delve into the next evolution of control flow management within Angular templates. With the release of Angular version 17, a new paradigm emerges: the built-in control flow. Let's delve into the details of this exciting new feature and explore how it enhances the Angular development experience.

## Built-in control flows

Built-in control flows offer a more concise and declarative way to manage control flow logic directly within your component templates, eliminating the need for structural directives. Here are the new built-in control flow statements.

### Built-in if statement

The `@if` statement conditionally renders content based on a Boolean expression.

Let's consider the previous example of `*ngIf`:

```
<div *ngIf="items.length === 0; else itemsFound">
  <div>No items found</div>
```

```
</div>
<ng-template #itemsFound>
  <div>Items found</div>
</ng-template>
```

Using the new `@if` and `@else` statements, the example will now look like this:

```
@if (items.length === 0) {
  <div> No items found </div>
} @else {
<div> Items found </div>
}
```

As you may have noticed, there are differences in syntax between the two code blocks. The `@if` and `@else` statements replace the `*ngIf` directive and the `ng-template` element by providing a more intuitive and JavaScript-like syntax for handling conditional rendering within component templates. You can optionally use an `@else` statement to provide alternative content when the condition evaluates to false.

Furthermore, while `*ngIf` requires importing `CommonModule` to function properly, `@if` is a standalone statement that can be directly used within the template without any additional imports.

Additionally, the `@if` block may have one or more associated `@else` blocks. After an `@if` block, you can optionally chain any number of `@else if` blocks and one `@else` block as follows:

```
@if (age >= 18) {
  You are an adult.
} @else if (age >= 13) {
  You are a teenager.
} @else {
  You are a child.
}
```

### Built-in for-loop statement

The `@for` statement iterates over a collection of data and renders content for each item.

Let's take the previous `*ngFor` example again:

```
<ul>
  <li *ngFor="let product of products; trackBy:
    trackProduct">{{ product.name }}
  </li>
</ul>
```

Using the new @for statement, the example will look like this:

```
@for (product of products; track product.id) {
   {{ product.name }}
}
```

Replacing the previously optional trackBy function used with *ngFor is the track function within the @for statement. Both approaches serve the same core purpose, enabling Angular to efficiently track changes within your iterated lists by focusing on the unique identifier of each item rather than its position in the array.

> **Note**
>
> While trackBy was optional, its absence often led to performance issues. However, using track is now mandatory within @for loops, ensuring optimal rendering speed by default.

A significant advantage of track is its ease of use compared to trackBy. You can directly include an expression representing the unique identifier of each item within the template itself, eliminating the need for a separate trackBy method in your component class (trackProduct in the previous example). This streamlines your code and improves readability.

The transition to track is designed to be seamless for developers who have already implemented trackBy functions and wish to migrate without removing those methods. They can seamlessly retain the existing methods and simply update the template as follows:

```
@for (product of products; track trackProduct($index, product) {
   {{ product.name }}
}
```

This ensures backward compatibility and a smooth transition process.

In essence, track offers a mandatory and simplified approach to change tracking within @for loops, promoting optimal performance and a more concise syntax in your Angular applications.

> **Note**
>
> It is worth mentioning that the @for statement uses a new diffing algorithm and offers a more optimized implementation compared to *ngFor. This enhancement results in up to 90% faster runtime according to community framework benchmarks. For more information, refer to https://krausest.github.io/js-framework-benchmark/current.html.

Furthermore, the built-in @for loop has a shortcut to deal with empty collections, referred to as the optional @empty block:

```
@for (product of products; track product.id) {
  {{ product.name }}
} @empty {
  Empty list of products
}
```

The @empty block offers a convenient and efficient way to display informative messages or alternative content when no data is available. It promotes a better user experience and keeps your component logic well-organized.

We went into a bit of detail there, so to summarize, here are the key benefits of the new @for statement:

- The @for syntax offers a cleaner and more readable way to iterate over lists, display alternative content when no data is available, and define unique identifiers for the list items.
- By requiring track, @for guarantees efficient DOM updates, leading to a smoother user experience.
- The @for loop leverages a new, optimized diffing algorithm compared to *ngFor. This has led to significant performance improvements, as evidenced by community benchmarks.

In essence, the @for statement provides an all-around upgrade for iterating over collections in your Angular applications. It empowers developers with a cleaner, more performant, and more user-friendly way to manage data within templates.

### Built-in switch statement

The @switch statement selects content based on a matching expression.

Let's take the previous example of *ngSwitch:

```
<div [ngSwitch]="userRole">
  <admin-dashboard *ngSwitchCase="admin" >
    </admin-dashboard>
  <user-dashboard *ngSwitchCase="'user'" >
    </user-dashboard>
  <guest-dashboard *ngSwitchDefault >
    </guest-dashboard>
</div>
```

Using the new @switch statement, it will now look like this:

```
@switch (userRole) {
  @case ('admin') { <admin-dashboard/> }
```

```
  @case ('user') { <user-dashboard/> }
  @default { <guest-dashboard/> }
}
```

As you may have noticed, both `@switch` and `*ngSwitch` achieve conditional rendering in Angular templates. However, `@switch` offers a more concise and modern approach that aligns better with current JavaScript practices. This syntax is more intuitive and closer to standard JavaScript switch statements, making code easier to understand and maintain.

The `@default` block is optional and can be excluded. In the absence of a matching `@case` and if there's no `@default` block provided, nothing will be displayed.

## Including built-in control flows in our recipe app

Now that we've learned about the new built-in control flow, let's take advantage of it and update our template code with this new syntax.

The HTML code of our `RecipesListComponent.html` file uses the Angular structural directives `*ngIf` (used to conditionally render the data view when the `recipes$` Observable returns a value) and `*ngFor` (used to iterate over the list of recipes and render a card for each recipe). Here is the code snippet:

```
<div *ngIf="recipes$ | async as recipes" class="card">
  <p-dataView #dv [value]="recipes" [paginator]="true"
  [rows]="9" filterBy="name" layout="grid">
    <ng-template let-recipes pTemplate="gridItem">
      <div class="grid grid-nogutter">
        <div class="col-12" class="recipe-grid-item card"
        *ngFor="let recipe of recipes">
          /** Extra code here **/
        </div>
      </div>
    </ng-template>
  </p-dataView>
</div>
```

Now, let's update this code using the new built-in control flow:

```
@if (recipes$ | async; as recipes) {
  <div class="card">
    <p-dataView #dv [value]="recipes" [paginator]="true"
    [rows]="9" filterBy="name" layout="grid">
      <ng-template let-recipes pTemplate="gridItem">
        <div class="grid grid-nogutter">
          @for (recipe of recipes; track recipe.id) {
```

```
            <div class="col-12"
            class="recipe-grid-item card">
/** Extra code here **/
            </div>
                }
        </div>
      </ng-template>
    </p-dataView>
  </div>
}
```

We replaced *ngIf with @if to conditionally render the data view when the recipes$ Observable returns a value.

We also replaced *ngFor with @for to iterate over the list of recipes and render a card for each recipe. We included within the @for statement the track function, track recipe.id. The recipe's ID is the unique identifier of the recipe.

We now have a refreshed template that not only is more performant but also aligns seamlessly with the latest version of Angular.

Additionally, if you have existing projects, you can easily migrate them to leverage the new built-in flow syntax by using the following migration schematic:

```
ng generate @angular/core:control-flow
```

## Benefits of built-in control flow

There are several benefits to using Angular's built-in control flow syntax, as follows:

- *Improved readability*: The syntax aligns more closely with JavaScript, making the code easier to understand and maintain.

- *Reduced boilerplate*: You can remove the need for separate directive imports and properties.

- *Built-in availability*: No additional imports are required; the feature is readily available out of the box in your templates.

- *Enhanced type safety*: The compiler provides more robust type narrowing, resulting in improved type safety and error detection.

- *Performance improvements*: While performance improvements can vary depending on your application's structure and data size, the @for statement utilizes a more streamlined diffing algorithm compared to *ngFor. This can potentially lead to smoother rendering and a better user experience, especially when dealing with large or frequently updated lists.

In short, the built-in control flow syntax fosters a more intuitive, concise, and performant approach to writing Angular templates. It promotes code readability, reduces boilerplate, and offers enhanced type safety.

## Summary

In this chapter, we explored the classic and reactive patterns for fetching data. We learned about the imperative way in which to manage unsubscriptions and the reactive pattern. We explained some useful RxJS operators, and also shed light on the advantages of using the reactive pattern and learned about all the technical aspects around it. We also learned about standalone components, a new edition to Angular, as well as how to create them, and what their benefits are. Lastly, we delved into the new built-in control flow introduced in Angular 17, covering its various applications, syntax, and associated benefits.

Now that we have retrieved our data as RxJS streams, in the next chapters, let's start playing with those streams to react to user actions using RxJS streams and, consequently, build our `RecipesApp` application in a reactive way. In the next chapter, we will focus on the reactive patterns for error handling and the different strategies that are available.

# Handling Errors Reactively

Errors in programming happen all the time, and RxJS is no exception. Handling those errors is a crucial part of every application. As I always say to my students in every training session, implementing a process that only covers happy cases determines the failure of your application. However, in RxJS, there are a lot of error handling strategies that you need to learn in order to handle errors efficiently.

We will start by explaining the contract of the Observable in RxJS, which is crucial to understanding what comes after. Then, we will learn the different error handling patterns and the operators provided by RxJS for that purpose. Next, we will shed light on the different error handling strategies and the use case of every strategy. Finally, we will practice one of the error handling strategies in our recipe app.

In this chapter, we're going to cover the following main topics:

- Understanding the anatomy of an Observable contract
- Exploring error handling patterns and strategies
- Handling errors in our recipe app

## Technical requirements

This chapter assumes that you have a basic understanding of RxJS. The source code of this chapter (except the samples) is available at `https://github.com/PacktPublishing/Reactive-Patterns-with-RxJS-and-Angular-Signals-Second-Edition/tree/main/Chap04`.

Please also refer to the *Technical requirements* section in *Chapter 3, Fetching Data as Streams*.

## Understanding the anatomy of an Observable contract

Understanding the anatomy of an **Observable contract** is crucial in order to learn error handling patterns. Let's dig deep into the Observable execution timeline by exploring the marble diagram explained in *Chapter 1, Diving into the Reactive Paradigm*:

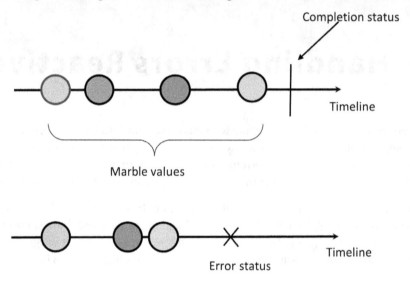

Figure 4.1 – The marble diagram elements

Let's examine the previous diagram. If we take a look at the stream's lifecycle, we can figure out that a stream has two final statuses:

- **Completion status**: Where the stream has ended without errors and will not emit any further values. It is a shutdown, i.e., the Observable completes.

- **Error status**: Where the stream has ended with an error and will not emit any further values after the error is thrown. It is also a shutdown.

Only one of those two states can occur, not both, and every stream can error out once. This is the Observable contract.

At this point, you may be wondering, How we can recover from an error then? This is what we will be learning in the following sections.

# Exploring error handling patterns and strategies

The first classic pattern we will learn for handling errors is based on the `subscribe()` method. The `subscribe()` method takes as input the object Observer, which has three callbacks:

- A **success callback**: This is called every time the stream emits a value and receives as input the value emitted

- An **error callback**: This is called when an error occurs and receives as input the error itself

- A **completion callback**: This is called when the stream completes

This is a basic example of a `subscribe` implementation:

```
stream$.subscribe({
    next: (value) => console.log('Value Emitted', value),
    error: (error) => console.log('Error Occurred', error),
    complete: () => console.log('Stream Completed'),
});
```

In the code sample, `stream$` represents our Observable, and we passed an object that has three callbacks to the `subscribe` method:

- A success callback that logs the received value in the console

- An error callback that logs the received error in the console

- A complete callback that logs the stream completion

So, in order to handle errors, the first possibility is implementing the error callback and tracing the error message, displaying an error popup to the user, or doing any other custom behavior. Pretty simple!

But wait! In *Chapter 3, Fetching Data as Streams*, we saw that we need to avoid the explicit `subscribe()` to streams and learned the reasons and limitations behind this, namely, that it is impossible to recover from the error or emit an alternative fallback.

That's right; in most cases, we will not be using `subscribe()` explicitly. I just wanted to show you the classic way to do this, which is not the best way. Instead, let's see some advanced error handling patterns and learn more operators that will help us in the error handling process.

I think you may be familiar with the try-catch statement available in many programming languages, which consists of a `try` block followed by one or more `catch` clauses. In the `try` block, you place your risky statements, and inside `catch`, you handle the possible exceptions:

```
try {
    // risky statements
}
catch(error) {
```

```
        // handle exceptions
    }
```

RxJS ships with the `catchError` operator, which provides us with something similar to the try-catch statement. The `catchError` operator is defined in the RxJS official documentation as an operator that "*catches errors on the Observable to be handled by returning a new Observable or throwing an error.*"

The `catchError` operator subscribes to the source Observable that might error out and emits values to the observer until an error occurs. When an error happens, the `catchError` operator executes a callback function, passing in the error. This callback function is responsible for handling errors and always returns an Observable.

If there are no errors, the output Observable returned by `catchError` works exactly the same way as the source Observable.

You can use `catchError` multiple times in an Observable chain, like so:

```
import { catchError} from 'rxjs/operators';

//stream$ is the source Observable that might error out
stream$.pipe(
    catchError(error => {
        //handle the error received
    })
).subscribe()
```

After calling the `catchError` operator, we need to implement the callback function that will handle the error. When it comes to handling errors, there are three strategies:

- The replace strategy
- The rethrow strategy
- The retry strategy

Let's break down these three strategies one by one in the following sections and explore some examples and use cases.

## The replace strategy

The **replace strategy** is named as such because the Observable returned by the error handling function will replace the Observable that has just errored out. This replacement Observable is then subscribed to, and its values are used instead of the errored-out input Observable. The following code is an example of this:

```
import { from, of } from 'rxjs';
import { catchError, map } from 'rxjs/operators';

const stream$ = from(['5', '10', '6', 'Hello', '2']);
stream$
  .pipe(
    map((value) => {
      if (isNaN(value as any)) {
        throw new Error('This is not a number');
      }
      return parseInt(value);
    }),
    catchError((error) => {
      console.log('Caught Error', error);
      return of();
    })
  )
  .subscribe({
    next: (res) => console.log('Value Emitted', res),
    error: (err) => console.log('Error Occurred', err),
    complete: () => console.log('Stream Completed'),
  });
```

```
//output
Value Emitted 5
Value Emitted 10
Value Emitted 6
Caught Error Error: This is not a number
Stream Completed
```

Let's break down what is happening in this example.

First, we have an Observable, `stream$`, created from an array of string values, `['5', '10', '6', 'Hello', '2']`, using the `from` creation operator. This operator creates an Observable that, when subscribing to it, will emit the array's values one by one and then complete.

> **Note**
>
> For more details about the `from` operator, please refer to the official documentation: `https://rxjs.dev/api/index/function/from#description`.

Next, we combined two operators in the pipe method of `stream$`:

- The `map` operator: This is used to transform the string values emitted to integers using the `parseInt()` method. If the value emitted is not a number, then an error is thrown with a `"This is not a number"` message.

- The `catchError` operator: We pass the error handling function to it, which will log the caught error and return `of()`. `of()` creates an Observable that has no values to emit, so it will immediately complete.

Then, we subscribe to the `stream$` and log a custom message in every callback to see what exactly happens at execution time.

At execution time, `stream$` will emit the string values of the array one by one (`'5'`, `'10'`, and `'6'`, respectively). The `map` takes those values one by one as input and returns 5, 10, and 6, respectively. `catchError()` takes the values emitted from the map operator and forwards them as output; the error handling function will not get called, as there is no error. Hence, the subscribers will receive 5, 10, and 6.

The `catchError()` operator comes into play when the `'Hello'` value is emitted. The map operator will throw an error, and the error handling function in `catchError()` will, consequently, get called. The error handling function, in our case, simply logs an error in the console and returns an Observable (created by the `of()` operator) that will immediately complete. This Observable will replace the current Observable that had an error; that's why we call it the replacement Observable.

`catchError()` will subscribe under the hood to the returned Observable. The `of()` Observable will complete immediately. Then, `stream$` is completed, so the next value, `'2'`, will not get emitted.

As you may have noticed, the error callback in the `subscribe()` method will not get called because we handled it in `catchError`. I added it on purpose to understand the behavior of error handling with `catchError`. Therefore, when an error occurs, the current stream that had an error out will get replaced by the stream returned from the `catchError()`; the values of the replaced Observable will then get emitted instead of the original stream values. This is what we call **fallback values**.

So, to summarize, the replace strategy is useful when we want to handle the error inside the stream itself and don't want the error to get propagated to the subscribers.

## The rethrow strategy

The **rethrow strategy** consists of rethrowing the error or, in other words, propagating the error to the subscribers of the output Observable of `catchError`. Notifying the subscribers about the error will help them perform side effects, such as displaying an error message in a popup.

To understand more about this strategy, let's look at the following example; it is the same as the one in the previous section, with the only difference being the error handling function:

```
import { from, throwError } from 'rxjs';
import { catchError, map } from 'rxjs/operators';

const stream$ = from(['5', '10', '6', 'Hello', '2']);
stream$
  .pipe(
    map((value) => {
      if (isNaN(value as any)) {
        throw new Error('This is not a number');
      }
      return parseInt(value);
    }),
    catchError((error) => {
      console.log('Caught Error', error);
      return throwError(() => error);
    })
  )
  .subscribe({
    next: (res) => console.log('Value Emitted', res),
    error: (err) => console.log('Error Occurred', err),
    complete: () => console.log('Stream Completed'),
  });
```

```
//output
Value Emitted 5
Value Emitted 10
Value Emitted 6
Caught Error Error: This is not a number
Error Occurred Error: This is not a number
```

In the error handling function, we return an Observable that is created using the `throwError` operator. The `throwError` operator creates an Observable that never emits any value; instead, it errors out immediately using the same error caught by `catchError`. In this way, the error will get pushed to the subscribers and can be further handled by the rest of the Observable chain if needed.

As you may have noticed, the same error was logged both in the `catchError` block and the subscriber error handler function, as expected, so the rethrow strategy has worked.

Please note that in the previous examples, we simply log the error in the console for demonstration purposes. However, in a real-world scenario, you can do much more, such as showing messages to the users.

## The retrying strategy

With the **retry strategy**, you can retry the Observable using the `retry` operator to give another chance to the stream. The `retry` operator retries an Observable a specific number of times and is useful for retrying HTTP requests or connections. We can see an example here:

```
import { catchError, map, retry } from 'rxjs/operators';
import { from, throwError } from 'rxjs';

const stream$ = from(['5', '10', '6', 'Hello', '2']);
stream$
  .pipe(
    map((value) => {
      if (isNaN(value as any)) {
        throw new Error('This is not a number');
      }
      return parseInt(value);
    }),
    retry(2),
    catchError((error) => {
      console.log('Caught Error', error);
      return throwError(() => error);
    })
  )
  .subscribe({
    next: (res) => console.log('Value Emitted', res),
    error: (err) => console.log('Error Occurred', err),
    complete: () => console.log('Stream Completed'),
  });


//output
Value Emitted 5
10
6
5
10
```

```
6
5
10
6

Caught Error Error: This is not a number
Error Occurred Error: This is not a number
```

As you may have noticed, the values of the source stream were emitted two times since we called the `retry` operator with 2 as a parameter; we gave the Observable two chances before throwing the error.

Now, in this case, we are retrying immediately. However, what if we want to retry in only specific cases or wait for a delay before retrying? This is where the `retryWhen` operator comes into play!

To understand the `retryWhen` operator, there's nothing better than a marble diagram:

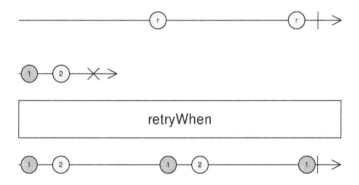

Figure 4.2 – The retryWhen operator

Let's explain what's going on here:

- The Observable in the first line is the **notifier Observable** that is going to determine when the retry should occur
- The Observable in the second line is the **source Observable** that will error out after emitting **1** and **2**

When we subscribe to the source Observable it will emit **1** and **2**. The `retryWhen` forwards those values as output. Then, the source Observable errors out and completes.

Nothing will happen until the notifier Observable emits the first value, **r**. At that moment, the source Observable will get retried, and as you see, **1** and **2** get emitted again. In fact, `retryWhen` will subscribe to the source Observable because it is already completed, so even if it is completed, it can be retried.

The notifier Observable is then going to emit another **r** value, and the same thing occurs.

Next, `retryWhen` starts to emit the first **1** value, but soon after, the notifier Observable completes; that's why the **2** value will not get emitted.

As you may have guessed, `retryWhen` retries the source Observable each time the notifier emits a value! This means that you can use this notifier Observable to emit values at the moment you want your source Observable to get retried and complete it at the moment you want your retry attempts to stop.

Now, let's have a look at the signature of the `retryWhen` operator:

```
export declare function retryWhen<T>(notifier: (errors:
Observable<any>) => Observable<any>):
MonoTypeOperatorFunction<T>;
```

The `notifier` parameter represents the callback that returns the notifier Observable and gets the error Observable as the argument. The error Observable will emit every time the source Observable errors out. So, `retryWhen` will subscribe to the notifier Observable and behave as described previously.

Here is the same example given in the replace and rethrow strategies, but using `retryWhen` instead:

```
import { from} from 'rxjs';
import { map, retryWhen, tap } from 'rxjs/operators';

const stream$ = from(['5', '10', '6', 'Hello', '2']);
stream$
  .pipe(
    map((value) => {
      if (isNaN(value as any)) {
        throw new Error('This is not a number');
      }
      return parseInt(value);
    }),
    retryWhen((errors) => {
      return errors.pipe(
        tap(() => console.log('Retrying the source
                              Observable...'))
      );
    })
  )
  .subscribe({
    next: (res) => console.log('Value Emitted', res),
    error: (err) => console.log('Error Occurred', err),
    complete: () => console.log('Stream Completed'),
  });


//Code runs infinitely
```

In the previous code, the first error is thrown when receiving the value `'Hello'`, which is not a number. The `retryWhen` operator will catch this error and get executed. Then, the notifier callback (the argument of `retryWhen`) simply takes the error Observable as input and returns it.

We also used the pipe to call the `tap` operator in order to log a message in the console (`'Retrying the source Observable...'`). The `tap()` operator is used to perform a side effect for each emitted value.

> **Note**
>
> For more details about the `tap` operator, please refer to this link from the official documentation: `https://rxjs.dev/api/operators/tap`.

If you execute that code, you will find out that it runs infinitely. Why? Because the source will always error out, and `retryWhen` will, consequently, subscribe infinitely to the source Observable.

If the source always errors out, it is not correct to retry immediately. However, the error will not always occur, for example, in the case of HTTP requests. Sometimes, the HTTP request fails because the server is down, or there is another temporary reason that may disappear, and the request might go through in the next attempts without any problem.

In that case, you can use the immediate retry or even a delayed retry, which retries after a certain delay, where we can wait, for example, for 5 seconds after the error occurs before retrying. That's what we will be learning in the next section.

Now, let's have a look at another operator that will help us implement the retry strategy: the `delayWhen` operator. The `delayWhen()` operator is used to delay values emitted from the source Observable by a given duration. It is similar to the `delay()` operator, but the delay duration is determined by an input Observable.

For more detail, let's take a look at a marble diagram:

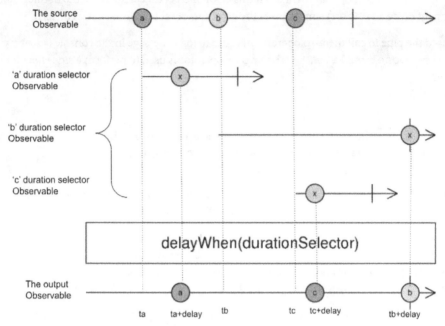

Figure 4.3 – The delayWhen operator

The first Observable is the source Observable. Each of the values, **a**, **b**, and **c**, has its own duration selector Observable, respectively, in the diagram: the **a** duration selector Observable, the **b** duration selector Observable, and the **c** duration selector Observable, which will emit one value, **x**, and then complete.

Every value emitted by the source Observable will be delayed before being emitted to the output Observable. In fact, when the source Observable emits the value **a** at **ta** in the timeline, the delayWhen operator will not immediately emit the value to the source Observable; instead, it will wait for the **a** duration selector Observable to emit a value at **ta+delay**, and at that exact time, the value **a** will get emitted to the Output Observable.

This carries on for the other values; the **b** value will show up in the Output Observable at **tb+delay** when the **b** duration selector emits a value, and the value **c** will get emitted at **tc+delay** when the **c** duration selector Observable emits a value. Note that here, **tb** and **tc** represent the emission time of the values **a** and **b** by the source Observable, respectively.

As you may have noticed, the value **b** was emitted before **c** by the source Observable (as **tb** precedes **tc**); however, the value **b** was shown after the value **c** in the output Observable (as **tb+delay** succeeds **tc+delay**); that's because the selector of **b** (the **b** duration selector in *Figure 4.3*) is emitted after the selector of **c** (the **c** duration selector also shown in *Figure 4.3*).

So, as you can see, the delay is completely flexible through the `durationSelector` function.

Another function, the `timer` function, can be useful in the delayed retry strategy:

```
export declare function timer(dueTime?: number | Date,
periodOrScheduler?: number | SchedulerLike, scheduler?:
SchedulerLike): Observable<number>;
```

This `timer` function returns an Observable and takes two arguments:

- `due`: A time period or exact date before which no values will be emitted

- `scheduler`: A periodic interval, in case we want to emit new values periodically

An example is `timer(5000,1000)`. The first value of the returned Observable will get emitted after 5 seconds, and a new value is emitted each second. The second argument is optional, which means that `timer(5000)` will emit a value after 5 seconds and then complete.

Now, it is time to combine the `delayWhen` and `retryWhen` operators to see how we can retry a failing HTTP request 5 seconds after each error:

```
import { Injectable } from '@angular/core';
import { HttpClient } from '@angular/common/http';
import { Recipe } from '../model/recipe.model';
import { catchError, delayWhen, of, retryWhen, tap, timer } from
'rxjs';
@Injectable({
  providedIn: 'root'
})

export class RecipesService {

recipes$ =
this.http.get<Recipe[]>('http://localhost:3001/recipes')
.pipe(
        retryWhen(errors => {
          return errors
            .pipe(
              delayWhen(() => timer(5000)),
              tap(() => console.log('Retrying the HTTP
                                    request...'))
            );
        }),
    );
constructor(private http: HttpClient) { }
}
```

> **Note**
>
> You will not find the preceding code in the GitHub repository, as it serves merely as an illustrative example. However, you can copy and paste it into the `RecipesService` class in order to test the delayed retry. Additionally, remember to stop the `recipes-book-api` mocked server to simulate retry attempts.

Our source Observable, in this case, is the result of an HTTP `get` request. Each time the request fails, the `delayWhen` operator creates a duration selector Observable through the `timer` function. This duration selector Observable is going to emit the 0 value after 5 seconds and then complete. Therefore, the notifier Observable of `retryWhen` will emit a value, and at that moment, the source Observable will get retried, after 5 seconds to be exact.

When you open the console, you will see this output:

```
⊗ ▶ GET http://localhost:4200/api/recipes 504 (Gateway Timeout)
    Retrying the HTTP request...
⊗ ▶ GET http://localhost:4200/api/recipes 504 (Gateway Timeout)
    Retrying the HTTP request...
⊗ ▶ GET http://localhost:4200/api/recipes 504 (Gateway Timeout)
    Retrying the HTTP request...
⊗ ▶ GET http://localhost:4200/api/recipes 504 (Gateway Timeout)
    Retrying the HTTP request...
⊗ ▶ GET http://localhost:4200/api/recipes 504 (Gateway Timeout)
>  |
```

Figure 4.4 – The failing HTTP request

As you may have noticed, every time the GET HTTP request fails, it is retried again after 5 seconds. That's how we achieved a delayed retry! So, to wrap up, each of the error handling strategies has its own techniques and serves a different purpose. In the following section, we will explore when to use each strategy.

## Choosing the right error handling strategy

Choosing the most appropriate error handling strategy in RxJS depends on various factors, such as the nature of the application, the type of errors encountered, and the desired user experience. Here's some guidance on when to use each strategy.

The replace strategy involves replacing the error with a fallback value or Observable. It's suitable in the following scenarios:

- You have a predefined fallback value or behavior to use when an error occurs, such as displaying placeholder content or default settings. For example, in a weather application, if fetching current weather data fails, you can replace the error with a default weather forecast for the user's location.

- The error is recoverable and doesn't require immediate intervention from the user.

- You want to provide a seamless user experience by gracefully handling errors without disrupting the application flow.

The rethrow strategy involves rethrowing the error to propagate it to the subscriber for handling. It's suitable in the following scenarios:

- You want to delegate error-handling responsibility to the subscriber or consumer of the Observable. For example, in an authentication service, if login fails due to invalid credentials, you can rethrow the error to allow the UI component to display an error message to the user.

- The error requires specific handling logic or customization based on the context in which it occurs.

- You want to provide flexibility for different parts of the application to handle errors differently.

The retry strategy involves retrying the operation that resulted in the error a certain number of times. It's suitable in the following scenarios:

- The error is transient or intermittent, such as network errors or temporary service disruptions where retrying the operation may succeed after subsequent attempts. For example, in a file upload service, if uploading a file fails due to a network error, you can retry the upload operation multiple times before giving up to ensure the file is successfully uploaded.

- Retrying the operation has a reasonable chance of success and can mitigate the impact of transient failures.

- You want to improve the reliability and robustness of operations that are prone to occasional failures.

Additionally, consider the following factors when choosing an error handling strategy:

- **User experience**: Consider how each strategy affects the user experience, such as whether it leads to delays, retries, or fallbacks.

- **Application requirements**: Align the chosen strategy with the specific requirements and constraints of your application, such as reliability, responsiveness, and error tolerance.

- **Performance implications**: Retry strategies may introduce additional overhead, especially if the operation involves expensive or time-consuming tasks.

Ultimately, the most appropriate error handling strategy depends on the specific context and requirements of your application. It's often beneficial to experiment with different strategies and observe their effects in real-world scenarios to determine the optimal approach.

Now that we have learned the different strategies and operators to handle errors, let's practice in our Recipes Book app in the next section.

## Handling errors in our recipe app

The first thing we are going to do is stop our mock service. Yes, you heard it right; stop it. This way, the call to the getRecipes service will fail because the server is down.

Now, if you refresh the front app, you will see that nothing, including our list of recipes, is displayed. Why do we get this behavior? Because we did not handle the errors. The error was thrown, the stream was completed, and nothing happened afterward. We have a white screen where nothing is displayed.

Now open the console, and you will see the failed request:

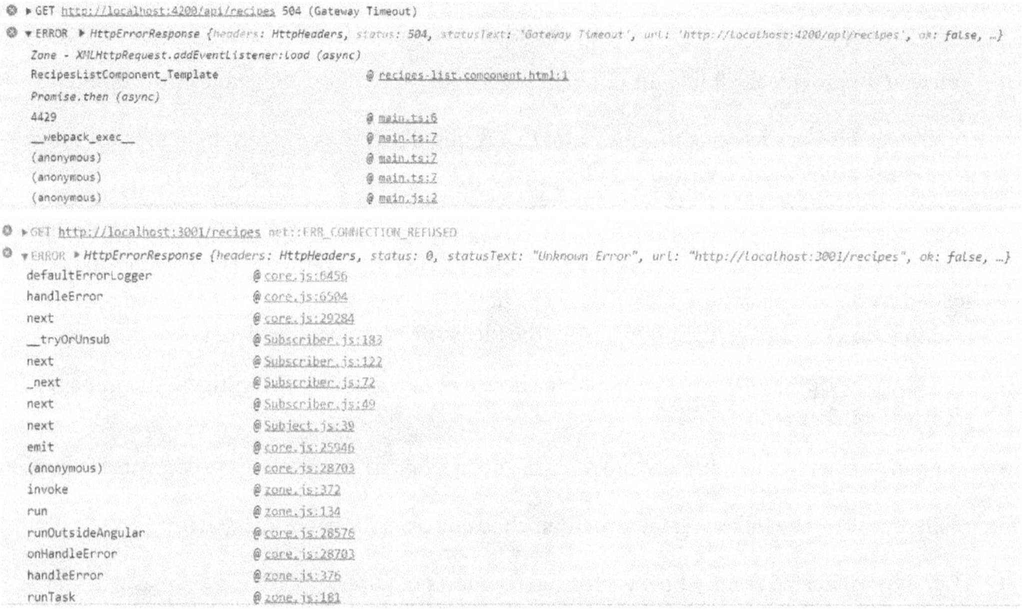

Figure 4.5 – The console showing the failed request

A failing HTTP request would never have broken our app if it was handled correctly. That's why you should be very careful when raising HTTP requests in your front application.

So, how can we fix this? Which strategy that we've previously discussed will fit the best?

If we choose the rethrow or retry strategy, then we will block the display of the recipes list. The user will get a blank page and will have to wait for the request to get executed successfully in order to see the list of recipes rendered in the screen. This is a valid option when you handle processes in the background that are not related to the UI display; however, if you raise requests in order to get results and display them in your UI components, then you should provide a replacement for that data to continue rendering the page. The user interface should keep on working regardless of whether or not there is an error; if there is an error, then we will display an empty list; if not, we will display the returned list from the server.

That's why the replacement strategy fits the most in this particular case. In fact, we want to get the list of recipes from the service, but if the service fails for whatever reason, I don't want my application to be frozen; I want to see a collection of zero elements (no elements), an empty table, or a list, and that is all. So, what we are going to do is use catchError and return an empty Observable, which is our fallback value.

Our service will look like this:

```
import { Injectable } from '@angular/core';
import { HttpClient } from '@angular/common/http';
import { Recipe } from '../model/recipe.model';
import { environment } from 'src/environments/environment';
import { catchError, of } from 'rxjs';
const BASE_PATH = environment.basePath

@Injectable({
  providedIn: 'root'
})

export class RecipesService {

  recipes$ = this.http.get<Recipe[]>(
    `${BASE_PATH}/recipes`).pipe(
      catchError(()=> of([])));

  constructor(private http: HttpClient) { }
}
```

This approach ensures that your application remains functional, displaying an empty list if you access the app. Moreover, you have the flexibility to customize the user interface by incorporating a message such as **There are no recipes**. To implement this, we'll make adjustments to our recipes-list. component.html as follows:

```
@if (recipes$ | async; as recipes) {
<div class="card">
```

```
<div>{{recipes.length}} Results</div>
<p-dataView #dv [value]="recipes" [paginator]="true"
[rows]="9" filterBy="name" layout="grid">
    <ng-template let-recipes pTemplate="gridItem">
        <div class="grid grid-nogutter">
            @for (recipe of recipes; track recipe.id) {
                <div class="col-12" class="recipe-grid-
                    item card">
                    /** Extra code here **/

                </div>
            } @empty {
                <div>There are no recipes</div>
            }
        </div>
    </ng-template>
</p-dataView>
</div>
} @else {
    <div>There are no recipes</div>
}
```

As you may have noticed, we used the new built-in control flow mechanism explained in *Chapter 3, Fetching Data as Streams*. By employing the @else block to the first if condition, we're able to display the message when no value is emitted from the recipes$ Observable. Additionally, the @empty statement added to the @for statement allows us to show the same message when the list of recipes is empty.

## Summary

In this chapter, we learned about the Observable contract and explored some of the most commonly used RxJS error handling strategies available and the different operators, namely catchError(), delayWhen(), retry(), and retryWhen(). We also shed light on the different strategies for error handling and when to choose each strategy. Finally, we handled the error in our Recipes Book app for the first implemented feature.

Now that we know how to handle errors in RxJS, let's move on to the next reactive pattern: combining streams.

# 5

# Combining Streams

So far, we have learned about the reactive pattern to fetch data as streams and have covered error-handling patterns. However, we have only explored the asynchronous data emitted from only one stream. What if we want to work with the asynchronous data emitted from different streams? Do you know how we can proceed?

Luckily, RxJS ships with one of the most powerful concepts: combining streams. Combining streams is the process of bringing together the emissions of multiple Observables in one stream. This allows you to explore multiple sources of asynchronous data as if they were a single stream. The main idea behind combining streams is manipulating asynchronous data in a more structured way.

This chapter revolves around a common use case, which is filtering data; we will resolve this by combining streams. We will start by explaining the filtering requirement, and then we will explore the imperative, classic pattern that can be used to implement this requirement, followed by a declarative, reactive implementation. Finally, we will highlight the common pitfalls to avoid when combining streams and discuss best practices.

In this chapter, we're going to cover the following main topics:

- Defining the filtering requirement
- Exploring the imperative pattern for filtering data
- Exploring the declarative pattern for filtering data
- Highlighting common pitfalls and best practices

## Technical requirements

The source code of this chapter (except the samples) is available at `https://github.com/ PacktPublishing/Reactive-Patterns-with-RxJS-and-Angular-Signals- Second-Edition/tree/main/Chap05`.

# Defining the filtering requirement

In our recipe application, we want to filter the displayed recipes according to certain criteria to refine the results. The following figure shows the implementation of the mockup described in the *View 1 – The landing page* section of *Chapter 2, Walking Through our Application*:

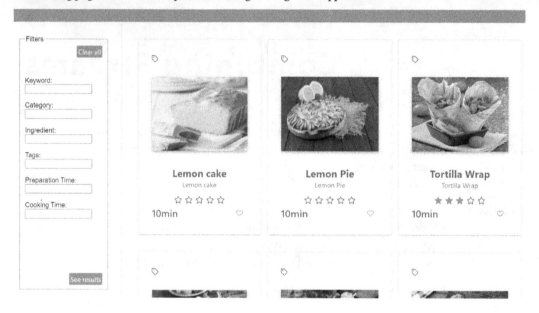

Figure 5.1 – The filtering requirement

From a user's perspective, the user will fill out some criteria in the **Filters** area and click on the **See results** button to see the results that match the filled criteria. The user can filter by using the keywords in the recipe title, recipe category, ingredients, tags, preparation, and cooking time. A filtering functionality is a must in the majority of applications that display data collections.

When it comes to filtering, there are a lot of strategies you can adopt, and the choice depends highly on the size of your data:

- If you have a small volume of data that you can fetch entirely on the client side, then it is unnecessary to perform server-side filtering; instead, it is faster to perform client-side filtering, which will not harm your application's performance.

- If you have a large amount of data, then you should be lazy when loading your data through pagination or virtual scroll in order to enhance the performance and the user experience as well. Therefore, in this case, conducting server-side filtering is unavoidable since you don't have all the data on the client side.

For demonstration purposes, we will filter only 11 recipes in total. So we're going to use client-side filtering (however, note that the reactive pattern we're going to discuss does not need a specific type of filtering – it can be used with both client and server-side filtering).

## Exploring the imperative pattern for filtering data

In this section, we will explore the imperative way to approach filtering data from the UI to the logic of filtering.

Let's start by setting up the UI code by creating a new standalone component, RecipesFilterComponent, under src/app. It is responsible for displaying the different filters to refine the initial results. The HTML template code of the filter component looks like this:

```
<div class="rp-data-view">
    <form [formGroup]="recipeForm">
        <fieldset class="rp-filters-group">
            <legend>Filters</legend>
            <div class="rp-filter-button">
                <p-button (onClick)="clearFilter()"
                label="Clear all"></p-button>
            </div>
            <label for="title">Keyword:</label>
            <input type="text" id="title"
            formControlName="title">
            <label for="category">Category:</label>
            <input type="text" id="category"
            formControlName="category">
            <label for="ingredient">Ingredient:</label>
            <input type="text" id="ingredient"
            formControlName="ingredient">
            <label for="text">Tags:</label>
            <input type="text" id="tags"
            formControlName="tags">
            <label for="text">Preparation Time:</label>
            <input type="text" id="prepTime"
            formControlName="prepTime">
            <label for="text">Cooking Time:</label>
            <input type="text" id="cookingTime"
            formControlName="cookingTime">
            <div class="rp-filter-button">
                <p-button class="rp-filter-button"
                (onClick)="filterResults()" label="See
                results"></p-button>
            </div>
```

```
        </fieldset>
    </form>
</div>
```

In the previous code, we used Angular reactive forms to display the search criteria inside a form. Then, we included two buttons: the **Clear all** button to clear the filters and the **See results** button to refine the displayed items by calling `filterResults()` in the `OnClick` callback. This method will replace the displayed recipes with those that match the filled criteria. Then, the UI will be updated automatically, thanks to the Angular change detection mechanism.

> **Note**
>
> For more details about Reactive forms, please refer to `https://angular.dev/guide/forms/reactive-forms`.

Now, let's move to `RecipesListComponent`, where we should consider a small change in the template. We should bind the `[value]` input of the `p-dataView` component to a new property, the `filteredRecipes` property, that holds the filtered results. The `filteredRecipes` property initially holds all the recipes requested from the server. This is what the template looks like:

```
<div class="card">
    <p-dataView #dv [value]="filteredRecipes"
    [paginator]="true" [rows]="9" filterBy="name"
    layout="grid">
    /** Extra code here **/
    </p-dataView>
</div>
```

The focus is not the HTML template, of course, but the process of filtering; however, it is important to point out when providing a complete workflow.

Now let's look at the logic of the classic pattern in `RecipesListComponent`:

```
export class RecipesListComponent implements OnInit,
OnDestroy {

  filteredRecipes: Recipe[] = [];
  recipes: Recipe[] = [];
  private destroy$: Subject<boolean> = new
  Subject<boolean>();

  constructor(private service: RecipesService, private fb:
  FormBuilder) {
  }
```

```
ngOnInit(): void {
  this.service.recipes$.pipe(takeUntil(this.destroy$))
    .subscribe((recipes) => {
      this.recipes = recipes;
      this.filteredRecipes = recipes;
    });
}

ngOnDestroy(): void {
  this.destroy$.next(true);
  this.destroy$.unsubscribe();
}


filterResults(recipe:Recipe) {
  this.filteredRecipes = this.recipes.filter(recipe =>
  recipe.title?.indexOf(recipe.title) !=
  -1)
}
```

Let's break down what is happening. Here, we declared three variables:

- `filteredRecipes`: The array that contains the filtered recipes. It is the property binded in the HTML template to the `[value]` input of the `p-dataview` component.

- `recipes`: The initial array of recipes.

- `destroy$`: A subject to clean up the subscription.

Then, in the `ngOninit()` method, we called the recipes (`recipes$`) that represent our Observable recipes. After that, we subscribe to it and initialize the recipes and `filteredRecipes` arrays with the emitted array from `recipes$`.

As we don't use the async pipe in this example, we should clean up the subscription manually using the `takeUntil()` pattern or `takeUntilDestroyed()`, as explained in *Chapter 3, Fetching Data as Streams*.

The `filterResults(recipe:Recipe)` method called when clicking on the **See results** button filters the current recipes list and returns the recipes that match the filter. In this sample, we considered only the title when filtering the criteria. The criteria are retrieved from the method input value through `recipe.title;`, containing the value the user filled out in the title input.

> **Note**
>
> For demonstration purposes, in this imperative implementation example, we've chosen to display `RecipesFilterComponent` within `RecipesListComponent`. This approach involves sending an output event, which encapsulates the filter form's value available in `RecipesFilterComponent`, to `RecipesListComponent`. Subsequently, `RecipesListComponent` executes the `filterResult` method based on this input

If you use server-side filtering, in other words, if you have a service that handles filtering data according to a given criteria, then `filterResults()` should call a backend service and will look like this:

```
filterResults() {
    this.filteredRecipes =
    this.http.get<Recipe[]>(`${BASE_PATH}/recipes`,
    {params:{criteria:this.recipeForm.value}});

}
```

That's it! It works fine, and I would say this is probably how most people would approach implementing filtering in an Angular application. The imperative way is kind of the obvious way to do it.

However, you may have noticed that we are no longer able to take advantage of `recipes$` as a stream in the template, as explained in *Chapter 3*, *Fetching Data as Streams*. Plus, what if your recipes service emits new recipes? This will overwrite the active filter until the user clicks on **See results** again to update the list with the current filter. This can be handled imperatively, of course, but it is a shame to use Observables and not take advantage of the power of reactivity.

So, without further ado, let's explore a better way to implement the filtering requirement by using a declarative and reactive way using the fundamentals of combining streams.

Remember, I always really want to highlight the classic way and the reactive way straight after to enable a smooth transition from imperative to declarative.

## Exploring the declarative pattern for filtering data

You should think of everything as a stream; this is the golden rule!

We have `recipes$`, which is already our **data stream**, but what if we consider the click on the **See results** button as a stream as well? We will call it the **action stream** and consider it an asynchronous data flow; we don't know when it happens, but every time the user clicks on **See results**, the action stream should emit the value of the filter.

So, in total, we need two streams:

- *The data stream*: In our case, it is `recipes$`, which is defined in `RecipesListComponent`, which we created in *Chapter 3*, *Fetching Data as Streams*:

```
export class RecipesListComponent {
/*Define The data stream */
recipes$ = this.service.recipes$;
constructor(private service: RecipesService) { }


}
```

- *The action stream*: In our case, it's named `filterRecipesSubject`; it is responsible for emitting the latest value of the filter every time the user clicks on the **Filter results** button. We will create it in the `RecipesService` service (which we also created in *Chapter 3*, *Fetching Data as Streams* as follows:

```
/*Create The action stream */
Private filterRecipeSubject = new
BehaviorSubject<Recipe>({title:''});
/* Extract The readonly stream */
filterRecipesAction$ =
this.filterRecipeSubject.asObservable();
```

Now, let's explain the previous code block. Here, we created two attributes:

- A **private** `BehaviorSubject` named `filterRecipeSubject` to prevent the external parts of the code from emitting values in the stream, erroring out, or completing the stream. We initialized `filterRecipeSubject` with a default value—an empty object—defined in the constructor argument. We can use either `Subject` or `BehaviorSubject` to create our action stream:

  - `Subject` is a special type of Observable in that it enables multicasting. We will explore multicasting in detail in *Chapter 9*, *Demystifying Multicasting*. For now, keep in mind that a subject is an observer and an Observable at the same time, so you can use it to share values among observers.

  - `BehaviourSubject` is a special type of subject that requires an initial value and always retains the last value to emit it to new subscribers. In other words, if you have any subscribers coming late to the game, they will get the last value emitted by the stream. This will always give you a value when you subscribe. We will discuss why we used `BehaviourSubject` instead of `Subject` at the end of the chapter.

- A **public** read-only stream named `filterRecipeAction$` (created from `filterRecipeSubject` through the method `asObservable()`) so that the other parts of the app can subscribe to and get the data.

So, just to recap, the `RecipesService` service will now look like this after adding the previous code block:

```
export class RecipesService {

  recipes$ =
    this.http.get<Recipe[]>(`${BASE_PATH}/recipes`);
  private filterRecipeSubject = new
    BehaviorSubject<Recipe>({title: '' });
  filterRecipesAction$ =
    this.filterRecipeSubject.asObservable();
  constructor(private http: HttpClient) { }


}
```

Now, it is time to combine the streams. Both of the streams rely on each other; when `recipes$` emits a new value, the filter should stay active, and when the filter emits a new value, the recipes list should be updated accordingly.

What we are really trying to do is get information from both streams. Whenever you want to join information from multiple Observables, you should think of the RxJS combination strategy. Instead of getting the data from both streams separately, we can combine it to form a single new stream. RxJS has a set of combination operators to use in that matter. In the next section, we will explore one of the most used combination operators, which is the `combineLatest` operator.

## The combineLatest operator

The `combineLatest` operator will combine the latest values emitted by the input Observables. So, every time one of the Observables emits, `combineLatest` will emit the last emitted value from each. Does that make sense? If not, don't worry; we will detail it further by using a marble diagram as usual:

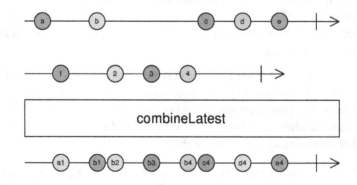

Figure 5.2 – The combineLatest marble diagram

Let's break this down:

- The first line in the marble diagram represents the timeline of the first input Observable.
- The second line represents the timeline of the second input Observable. So, combineLatest has two inputs.
- The last line represents the timeline of the output Observable returned from the combineLatest operator.

Now, let's dig deeper into the execution.

The first Observable emitted the value **a** first. At that time, the second Observable had not emitted anything, and nothing was emitted from combineLatest. Why? Because combineLatest will not emit until all the input Observables emit one value each. So, when the second Observable emits **1**, combineLatest will emit **a1**.

Bear in mind that combineLatest will not emit an initial value until each Observable emits at least one value.

Then, the first Observable emitted another value, **b**. At that time, the second Observable had not emitted anything, but its last emitted value was **1**, so combineLatest will emit the last value emitted by each input stream, which is **b1**. Then, the second Observable emitted is **2**. At that time, the latest value emitted by the first Observable was **b**, so combineLatest will emit **b2**, and so on and so forth.

Let's come back to our example and see how we can combine the data stream and the action stream that we have just created in order to filter results reactively using the combineLatest operator.

In RecipesListComponent, we will create a new stream named filteredRecipes$, which represents the result of combining the data and the action stream:

```
filterRecipesAction$ = this.service.filterRecipesAction$;
filteredRecipes$ = combineLatest([this.recipes$,
  this.filterRecipesAction$])
```

So, here, we used the combineLatest operator and passed to it (as a parameter) an array of two values: the first one is the recipes$ data stream, and the second one is the filterRecipeAction$ action stream. combineLatest will then return an array of two values (as the number of input Observables): the first element of the array is the last emitted value from the first stream, and the second element of the array is the last emitted value from the second stream. It respects the order.

Now, we will use filteredRecipes$ from the RecipesListComponent template, where we will bind it to the [value] input of p-dataview so that the following occurs:

- filteredRecipes$ should return all the recipes when loading the page
- filteredRecipes$ should return only the recipes that match the selected criteria when filtering

Then, the HTML code of the `RecipesListComponent` template will look like this:

```
@if (filteredRecipes$ | async; as recipes) {
<div class="card">
    <p-dataView #dv [value]="recipes" [paginator]="true"
    [rows]="9" filterBy="name" layout="grid">
        /** Extra code here **/
    </p-dataView>
</div>
} @else {
<div>There are no recipes</div>
}
```

> **Note**
>
> In the template, we have included both the `RecipesList` and `RecipesFilter` components. This separation of components enhances the maintainability and readability of the codebase, promoting a modular and scalable architecture for the application.

Still, there is one change that we need to make in `RecipesListComponent`. In our UI, we want to display all the recipes when loading the page and the filtered recipes when refining the results; because of this, we need to edit `filteredRecipes$` so it should not return the result of the combination directly (and in most cases, that won't happen anyway). Instead, we will process the result of the combination (the returned array), take the information that we need from the combined streams, and transform it into whatever we need.

In this case, we want to modify the result stream so that rather than just giving us the latest recipes and criteria values, it will give us an array of recipes filtered by the received criteria as follows:

```
filteredRecipes$ = combineLatest([this.recipes$,
  this.filterRecipesAction$]).pipe(
    map((resultAsArray: [Recipe[], Recipe]) => {
      const filterTitle =
        resultAsArray[1]?.title?.toLowerCase() ?? '';
      return resultAsArray[0].filter(recipe =>
        recipe.title?.toLowerCase().includes(filterTitle));
    })
  );
```

The result of the combination is the `resultAsArray` parameter. The first element, `resultAsArray [0]`, represents the last emitted recipes from `recipes$`, and the second element, `resultAsArray [1]`, represents the last emitted criteria from `filterRecipesAction$`.

However, we can do even better! We can enhance the code using the array destructuring technique like so:

```
filteredRecipes$ = combineLatest([this.recipes$,
  this.filterRecipesAction$]).pipe(
    map(([recipes, filter]: [Recipe[], Recipe]) => {
      const filterTitle =
        filter?.title?.toLowerCase() ?? '';
      return recipes.filter(recipe =>
        recipe.title?.toLowerCase().includes(filterTitle))
    })
  );
```

Here, the `recipes` parameter represents the first element of the returned array, and the `filter` parameter represents the second element of the returned array. After `:`, we have the parameter types, and that's it. So, instead of obtaining the elements by using the index directly, the array destructuring technique provides you with a way to name the array elements and get the values directly from those variables. Finally, we used the `filter` method to filter the list of recipes that match the criteria.

At this point, we put in place all the mechanisms to filter the values reactively. The last thing left to do is update the filter value every time it changes, and this is what we're going to explore in the next section.

## Updating the filter value

As we said, every time the user clicks on the **See results** button, `filterRecipesAction$` should emit the criteria so that our `combineLatest` re-executes and returns the filtered recipes. To achieve this, we created a new method called `updateFilter` in the `RecipesService` that takes the filter value as input and simply emits it using the next method over the `filterRecipesSubject` subject:

```
updateFilter(criteria:Recipe) {
  this.filterRecipeSubject.next(criteria);
}
```

Then, we will call this method in `RecipesFilterComponent` when the user clicks on the **See results** button to update the filter value:

```
filterResults() {
  this.service.updateFilter(<Recipe>this.recipeForm.value);
}
```

We passed the value of the "Filters" form created in the `RecipesFilterComponent` to the `updateFilter` method. That's it; now, on every emission of criteria, it will re-execute the `combineLatest` operator and will consequently filter the values.

To summarize, this is what the complete code looks like in `RecipesListComponent`:

```
@Component({
  selector: 'app-recipes-list',
  standalone: true,
  imports: [CommonModule],
  templateUrl: './recipes-list.component.html',
  styleUrls: ['./recipes-list.component.scss'],
  changeDetection: ChangeDetectionStrategy.OnPush
})
export class RecipesListComponent {

  /*The data stream */
  recipes$ = this.service.recipes$;
  filteredRecipes$ = combineLatest([this.recipes$,
  this.filterRecipesAction$]).pipe(
    map(([recipes, filter]: [Recipe[], Recipe]) => {
    const filterTitle = filter?.title?.toLowerCase() ?? '';
      return recipes.filter(recipe =>
      recipe.title?.toLowerCase().includes(filterTitle)
    })
  );

  constructor(private service: RecipesService) {
  }
}
```

The `RecipesFilterComponent` component looks like this:

```
@Component({
  selector: 'app-recipes-filter',
  standalone: true,
  imports: [ButtonModule, ReactiveFormsModule],
  templateUrl: './recipes-filter.component.html',
  styleUrl: './recipes-filter.component.css'
})
export class RecipesFilterComponent {
  recipeForm = this.fb.group<Recipe>({
    title: '',
    category: '',
    ingredients: '',
    tags: '',
    prepTime: undefined,
```

```
    cookingTime: undefined,
  });

constructor(private service: RecipesService, private fb:
  FormBuilder) { }

filterResults() {
this.service.updateFilter(<Recipe>this.recipeForm.value);
}
clearFilters() {
this.recipeForm.reset();
}
```

Finally, `RecipesService` looks like this:

```
export class RecipesService {

  recipes$ = this.http.get<Recipe[]>(
  `${BASE_PATH}/recipes`);
  private filterRecipeSubject = new
  BehaviorSubject<Recipe>({ title: '' });
  filterRecipesAction$ =
  this.filterRecipeSubject.asObservable();

  constructor(private http: HttpClient) { }

  updateFilter(criteria:Recipe) {
    this.filterRecipeSubject.next(criteria);
  }

}
```

---

**Note**

The complete code is available in the GitHub repository.

---

Now, let's answer the question, Why did we choose `BehaviorSubject` instead of `Subject`?

Between loading the page and clicking on **See results**, we need `filteredRecipes$` to hold all the recipes, as explained in the *Combine streams* section. If we use a plain `Subject`, the criteria will only get emitted when we click on the button. That means that when loading the page, only `recipes$` is emitted, and `combineLatest` will wait for all the streams to emit one value before emitting more. In our UI, we would then get an empty list.

However, when we use `BehaviorSubject`, it will emit the default value for all the subscribers immediately, so `combineLatest` will emit a first value, and everything will work fine and that's it. Seems like magic, right?

Here's an example of the filtered recipes when searching for `Lemon` in the keyword:

Figure 5.3 – Filtered recipes

To sum up, in order to resolve the use case we've looked at in this chapter, we started with defining the data stream responsible for getting the data; then, we created the action stream. Afterwards, we combined the streams, did some manipulation on the combined stream, binded it in our template, and, finally, updated the filter value in the filter event.

Now, let's highlight some of the common pitfalls when using `combineLatest`.

# Highlighting common pitfalls and best practices

Here are some common pitfalls or scenarios to avoid when using `combineLatest` in RxJS:

## Unnecessary subscriptions

`combineLatest` subscribes to all provided Observables automatically. Make sure you unsubscribe from the returned subscription when you no longer need the combined values, especially for long-running or infinite Observables. This prevents memory leaks and unnecessary processing.

## Missing or incomplete values

`combineLatest` only emits a value when all its source Observables have emitted at least one value. If any Observable completes or throws an error before all have emitted, `combineLatest` will also complete or error out, respectively. Consider using the `withLatestFrom` operator if you only need the latest value from one Observable combined with the entire emission history from another.

## Performance overhead

Combining a large number of Observables with `combineLatest` can introduce performance overhead. Evaluate the need for combining so many streams and consider using simpler operators, such as `forkJoin`, if you only need all emissions as a single array once all the Observables complete. We will delve into the `forkJoin` operator in *Chapter 11*, *Performing Bulk Operations*.

## Confusing error handling

`combineLatest` propagates errors from any of its source Observables. If error handling is complex, consider using custom operators to handle errors for each Observable individually before combining them. Implement proper error handling by using the `catchError` operator on individual Observables or using a custom operator that combines the emissions and errors from each source.

As you can see, `combineLatest` is a powerful operator for combining multiple Observables, but it's important to understand its behavior and potential pitfalls in order to use it effectively in your RxJS applications. Choose the right operator based on your specific use case and prioritize clear and maintainable code.

# Summary

In this chapter, we embarked on a journey through filtering data, beginning with an exploration of the imperative pattern. We then transitioned to one of the most commonly used RxJS patterns for data filtering, which can be used in many use cases where data updates are triggered by actions. By delving deeper, we outlined the different steps needed to implement the reactive pattern, from creating the streams to their combination, using the `combineLatest` operator. We learned how this operator works and how we can use it in a practical implementation. We used the combined stream in our template and handled data updates reactively. Finally, we explored some of the common pitfalls to avoid when using `combineLatest`.

Combining streams is a fundamental concept in RxJS that enables you to build more complex and powerful asynchronous data processing. It allows you to join data from various sources and apply operators and transformations to create new streams that meet your specific application requirements.

Now that we know about this useful pattern, let's move on to the next reactive pattern, transforming streams.

# Transforming Streams

When dealing with streams, one of the most frequent use cases you will face is the need to transform a stream of certain values into a stream of other values. That's what this chapter is about.

This chapter revolves around adding an autosave feature to our project that we will be resolving by transforming streams. We will start by explaining the autosave requirement that we will be implementing in the recipe app. Then, we will explore the imperative way of implementing this feature. After that, we will learn about the declarative pattern for doing it and study the most commonly used RxJS transformation operators for this situation.

Finally, we will delve into the different transformation operators provided by RxJS and their respective use cases, enriching our understanding through hands-on examples.

So, in this chapter, we're going to cover the following main topics:

- Defining the autosave requirement
- Exploring the imperative pattern for the autosave feature
- Exploring the declarative pattern for the autosave feature

## Technical requirements

This chapter assumes that you have a basic understanding of RxJS.

For more details about Reactive forms, please refer to `https://angular.dev/guide/forms/reactive-forms`.

For demonstration purposes, we will be using a fake autosave service. Its implementation is available in the `recipes-book-api` module in this book's GitHub repository. Note that we won't be going through the details of this service as the focus is not the backend of the project.

The source code of this chapter is available at `https://github.com/PacktPublishing/Reactive-Patterns-with-RxJS-and-Angular-Signals-Second-Edition/tree/main/Chap06`.

# Defining the autosave requirement

As described in the *View 2 – The New Recipe interface* section of *Chapter 2, Walking through Our Application*, the user can add a new recipe by clicking on the **New Recipe** menu item. This will display the following form to be filled out:

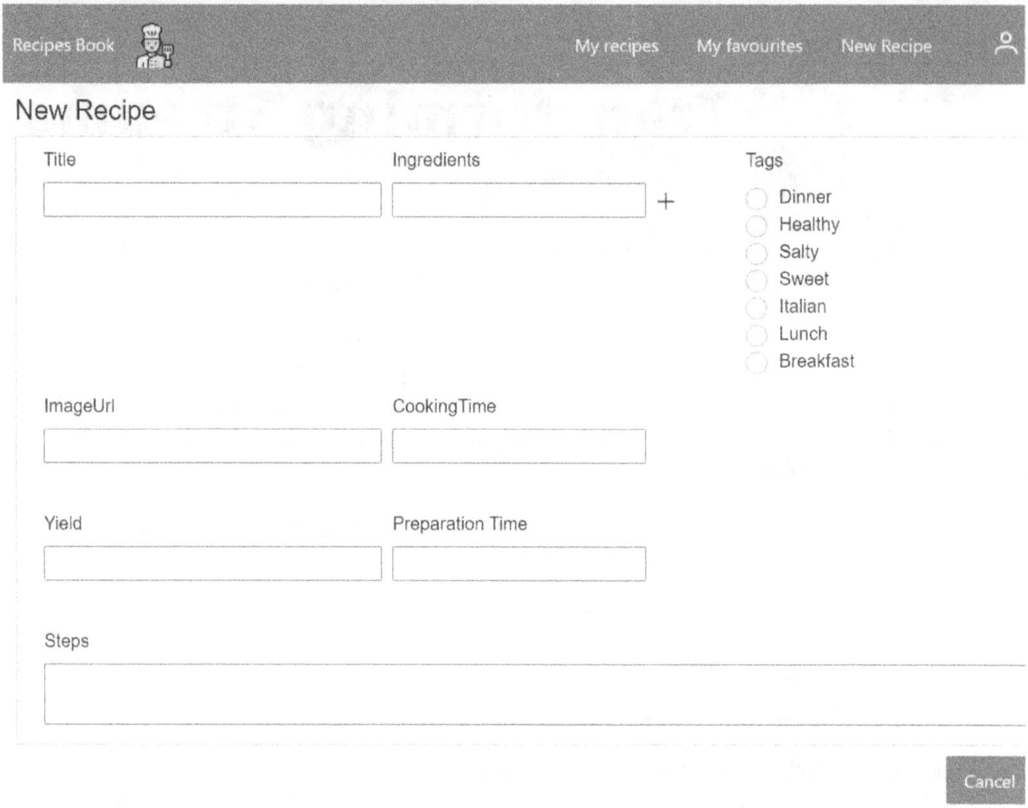

Figure 6.1 – The New Recipe form

The standalone component that's responsible for displaying the **New Recipe** form is called `RecipeCreationComponent` and is available under `recipes-book-front\src\app\ recipe-creation`.

Here, we want to implement the autosave behavior, which consists of storing the user's changes in the form automatically. In this example, we will be storing the form changes in the backend so that the user can retrieve the last changes any time after a disconnection, a timeout, or other problems – this feature improves the user experience by preventing data loss.

Now that we understand the requirement, let's look at the imperative way to implement the autosave feature.

# Exploring the imperative pattern for the autosave feature

We used Angular Reactive forms to build the **New Recipe** creation form. As described in the *Using RxJS in Angular and its advantages* section of *Chapter 1, Diving into the Reactive Paradigm*, Reactive forms leverage RxJS by providing the `valueChanges` Observable to track the `FormControl` changes. This makes our implementation easier since we want to listen to the form's value changes to perform a save on every change.

You can find the HTML code of the **New Recipe** creation form in the `recipe-creation.component.html` file template. Then, in `recipe-creation.component.ts`, we can define the form as follows:

```
export class RecipeCreationComponent implements OnInit {

  constructor(private formBuilder: FormBuilder) { }

  recipeForm = this.formBuilder.group<Recipe>({
    id: Math.floor(1000 + Math.random() * 9000),
    title: '',
    ingredients: '',
    tags: '',
    imageUrl: '',
    cookingTime: undefined,
    yield: 0,
    prepTime: undefined,
    steps: '',
  });
  tags = recipeTags.TAGS;
```

Here, we used the Angular `FormBuilder` API to build the reactive form and pass to it a JSON object where we define the different fields of the form. This JSON object represents our recipe's data; we're going to save this later. Every time we open the **New Recipe** form, a new empty object will be created.

Note that the first property of this JSON object, `id`, is not going to be displayed in the form. We only add it to initialize the new `Recipe` object with a random identifier to save the recipe's data properly in the backend. The `tags` property is retrieved from a constant declared in `src/app/core/model/tags.ts` that represents the static list of available tags.

Now that we've prepared our form, let's see how we can implement the autosave feature. The first thing that comes to mind is subscribing to the valueChanges Observable of recipeForm in the ngOninit() instance of RecipeCreationComponent. Then, every time the valueChanges Observable emits a new form value, we should raise a save request to save the most recent value of the form. We can do this like so:

```
ngOnInit(): void {
    this.recipeForm.valueChanges.subscribe(
      formValue => {
        this.service.saveRecipe(<Recipe>formValue);
      }
    );
```

The saveRecipe method is then defined and implemented in RecipeService, as follows:

```
saveRecipe(formValue: Recipe) : Observable<Recipe>  {
   return this.http.post<Recipe>(`${BASE_PATH}/recipes`,
      formValue);
}
```

Here, we use the HTTPClient API and call the save service in the backend.

> **Note**
>
> Backend implementation is not the focus of this book. For that reason, we've provided a fake implementation of the POST save service in the recipes-book-api project. Here, the goal is to simulate the call to an HTTP request to save the data.

So, to recap, the code for RecipeCreationComponent will look like this:

```
export class RecipeCreationComponent implements OnInit {

    constructor(private formBuilder: FormBuilder, private
    service: RecipesService) { }

    recipeForm = this.formBuilder.group<Recipe>({
      id: Math.floor(1000 + Math.random() * 9000),
      title: '',
      ingredients: '',
      tags: '',
      imageUrl: '',
      cookingTime: undefined,
      yield: 0,
      prepTime: undefined,
```

```
  steps: '',
});

tags = recipeTags.TAGS;


ngOnInit(): void {
  this.recipeForm.valueChanges
    .subscribe(
      formValue => {
        this.service.saveRecipe(<Recipe>formValue);
      }
    );


}

}
```

However, this code won't work. You should know by now that the result of this.service.
saveRecipe(<Recipe>formValue), which calls this.http.post<Recipe>(`${BASE_
PATH}/recipes`, formValue), is an Observable, and since Observables are lazy, we should
subscribe to this.service.saveRecipe(<Recipe>formValue) to initiate the HTTP
POST request. So, let's add a subscribe value, like so:

```
ngOnInit(): void {
    this.recipeForm.valueChanges.subscribe(
      formValue => {
        this.service.saveRecipe(<Recipe>formValue)
          .subscribe(
            result => this.saveSuccess(result),
            errors => this.handleErrors(errors)
          );
      }
    );
```

As you may have noticed, we called a subscribe value inside another subscribe, something
we call a nested subscription. However, this is considered an anti-pattern in RxJS and is problematic
for several reasons:

- Every time we use subscribe(), we open the door to imperative code. As we have learned
  throughout this book, we should avoid this as much as possible.

- Nested subscriptions require careful cleanup; otherwise, we can run into various performance problems. In the previous example, we didn't clean up the subscriptions, which means the possibility of serious timing issues arises. If multiple form values are emitted by `valueChanges` successively, many save requests will be sent in parallel. If the requests take some time to complete, there is no guarantee that the backend will process the save requests in order. For instance, we cannot ensure that the last valid form value is the one that's been saved in the backend. Consequently, we will end up with data incoherence.

What we want to do is perform a save request after the previous one is completed. Luckily, RxJS includes some interesting operators that fix this for us. So, without further ado, in the following section, we'll learn how to implement this in a reactive and declarative way.

## Exploring the reactive pattern for the autosave feature

You remember the golden rule from *Chapter 5*, right? We should think of everything as a stream. So, let's start by identifying our streams.

Here, we can think of the save operation as a stream – it is the result of the `this.service.saveRecipe(<Recipe>formValue)` method, which calls `this.http.post<Recipe>(`${BASE_PATH}/recipes`, formValue`. We will call it `saveRecipe$`.

The `saveRecipe$` Observable is responsible for saving the data in the backend and will initiate the `http` request when subscribed to.

To avoid nesting subscriptions, what we can do in this situation is map or transform the form value emitted by the `valueChanges` Observable to the `saveRecipe$` Observable. The result is what we call a higher-order Observable.

Not clear? Don't worry – we will explain this in detail in the next section.

### Higher-order Observables

So, what is a higher-order Observable? A **higher-order Observable** is just an Observable like any other, but its values are Observables as well. So, instead of emitting simple values such as strings, numbers, or arrays, it emits Observables that you can subscribe to separately.

Okay, but when is it useful? You can create a higher-order Observable whenever you use data emitted from one Observable to emit another Observable. In our case, for every emitted form value from the `valueChanges` Observable, we want to emit the `saveRecipe$` Observable. In other words, we want to transform (or map) the form value to the `saveRecipe$` Observable. This would create a higher-order Observable where each value represents a save request.

In this situation, the `valueChanges` Observable is called the outer Observable, and `saveRecipe$` is called the inner Observable. Under the hood, we want to subscribe to each `saveRecipe$` Observable that's emitted and receive the response all in one go to avoid nested treatments.

Now that we've learned what higher-order Observables are and when to use them, let's look at higher-order mapping operators.

## Higher-order mapping operators

To transform the outer Observable, we should use higher-order mapping operators. The role of these operators is to map each value from an outer Observable to a new inner Observable and automatically subscribe and unsubscribe to/from that inner Observable.

But what is the difference between regular mapping and higher-order mapping?

Well, regular mapping involves mapping one value to another value. One of the most used basic mapping operators is the `map` operator:

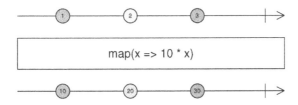

Figure 6.2 – The map operator – marble diagram

As described in this marble diagram, the map operator will transform the values of the input stream by multiplying each emitted value by 10. Here, $x=>10*x$ is the transformation function.

On the other hand, higher-order mapping is about mapping one value into an Observable.

RxJS provides several higher-order mapping operators. In the next section, we will learn about the `concatMap()` operator – which we will use to implement the autosave behavior – before discovering some other commonly used operators.

### The concatMap operator

`concatMap` is a combination of the concatenation strategy and higher-order mapping:

*concatMap = concat (concatenation) + map (higher-order mapping)*

We looked at the concepts of regular and higher-order mapping in the previous section, so let's look at the following marble diagram to understand the concatenation strategy, taking the example of the `concat` operator:

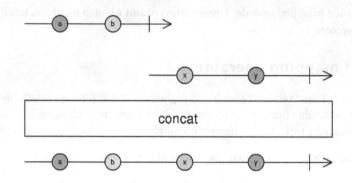

Figure 6.3 – The concat operator – marble diagram

Let's break down the marble diagram:

- The first line represents the timeline of the first Observable passed as input to the concat operator.

- The second line represents the timeline of the second Observable passed as input to the concat operator.

- The concat operator in this example has two inputs. It will subscribe to the first Observable but not to the second one. The first Observable will emit the values **a** and **b**, which get reflected in the result Observable (the last line).

- Then, the first Observable completes, and at that moment, the concat operator subscribes to the second Observable. This is how a sequential process is guaranteed.

- The second Observable will emit the values **x** and **y**, which get reflected in the result Observable.

- When the second Observable completes, the output Observable will also complete.

As you may have noticed, Observable concatenation is all about Observable completion. This is the key point. It emits the values of the first Observable, waits for it to complete, and then emits the values of the next Observable, and so on, until all the Observables complete.

Now that we understand the concatenation strategy, we can understand how the concatMap operator is a mixture of higher-order mapping and Observable concatenation: it waits for each inner Observable to complete before processing the next one. It's like a line at a ticket counter, where each customer (Observable) waits for their turn to be served before the next one is called.

## Using concatMap for autosaving

Based on the previous discussion, the `concatMap` operator fits very well with our autosave requirement for the following reasons:

- We want to take the form value and turn it into a `saveRecipe$` Observable and automatically subscribe and unsubscribe from the `saveRecipe$` inner Observable – this is what a higher-order mapping operation does.

- We only want to perform a save request after the previous one is completed. When one HTTP save request is in progress, the other requests that come in the meantime should wait for its completion before being called to ensure sequentiality. So, we need to concatenate the `saveRecipe$` Observables together.

This is what our code will look like:

```
valueChanges$ = this.recipeForm.valueChanges.pipe(
  concatMap(formValue =>
    this.service.saveRecipe(<Recipe>formValue)),
  catchError(errors => of(errors)),
  tap(result => this.saveSuccess(result))
);
```

Let's break down what's going on:

- Here, the outer Observable, `this.recipeForm.valueChanges`, emits form values. For each emitted form value, `concatMap` transforms it into `this.service.saveRecipe(<Recipe>formValue)`, which is the `saveRecipe$` Observable – our inner Observable.

- `concatMap` automatically subscribes to the inner Observable and the HTTP POST request will be issued.

- Another form value might come faster than the time it takes to save the previous form value in the backend. In this case, the form value will not be mapped to the `saveRecipe$` Observable. Instead, `concatMap` will wait for the previous save request to return a response and complete it before transforming the new form value to `saveRecipe$`, subscribing to it, and sending a new save request. When all inner Observables complete, the result stream completes.

- Then, we use the `catchError` operator to handle the errors and register a side effect with the `tap` operator to log the `Saved successfully` message in the backend. You can customize this, of course, and display a message to the end user.

To recap, the complete code for `RecipeCreationComponent` will now look like this:

```
export class RecipeCreationComponent {

  constructor(private formBuilder: FormBuilder, private
  service: RecipesService) { }

  recipeForm = this.formBuilder.group<Recipe>({
    id: Math.floor(1000 + Math.random() * 9000),
    title: '',
    ingredients: '',
    tags: '',
    imageUrl: '',
    cookingTime: undefined,
    yield: 0,
    prepTime: undefined,
    steps: '',
  });

  tags = recipeTags.TAGS;
  valueChanges$ = this.recipeForm.valueChanges.pipe(
    concatMap(formValue =>
      this.service.saveRecipe(<Recipe>formValue)),
    catchError(errors => of(errors)),
    tap(result => this.saveSuccess(result))
  );


  saveSuccess(_result: Recipe) {
    console.log('Saved successfully');
  }


}
```

Now, there's just one thing left to do: we should subscribe to the `valueChanges$` Observable to make all of this work. As usual, we will do this through the async pipe in our `RecipeCreationComponent` HTML template, as follows:

```
<ng-container *ngIf="valueChanges$ | async">
  </ng-container>
/** All the form code here**/
```

With that, the reactive implementation is complete.

As you may have noticed, the first benefit of using `concatMap` is that we no longer have nested subscriptions. We also get rid of explicit subscriptions thanks to the async pipe. Besides this, all form values are going to be sent to the backend sequentially.

When a delay is introduced in the backend save service (which I've set up by default in the provided implementation of the save recipe service), you'll notice that requests aren't initiated while another one is still processing. Instead, they wait until the current request finishes before being triggered. This is exactly what `concatMap` aims to achieve.

Now, let's take a look at the behavior `concatMap` manifests in the Chrome DevTools **Network** tab:

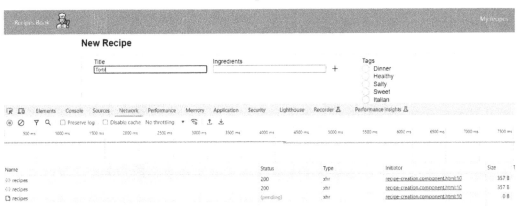

Figure 6.4 – The concatMap network requests

Here, upon typing a character, a POST save request is sent to the server. When trying to input other characters before the initial request, a response is received. You'll notice that requests are not immediately triggered; they are queued and executed sequentially once the preceding request is completed.

Our example served as a straightforward illustration of how the `concatMap` operator works. However, we could optimize our implementation more to avoid sending requests for every character introduced by the user. To do so, we could use the `debounceTime(waitingTime)` operator to wait for the user input to stabilize before sending the requests. We could also optimize it further by ignoring duplicates and making the `distinctUntilChanged()` operator handle invalid values.

> **Note**
>
> For more details about the `debounceTime` and `distinctUntilChanged` operators, refer to `https://rxjs.dev/api/operators/debounceTime` and `https://rxjs.dev/api/operators/distinctUntilChanged`, respectively.

## Using concatMap for pagination

Besides autosaving, we can also use `concatMap` for list pagination. In our recipe app, we handle the Recipes list pagination on the client side – we retrieve all the recipes when loading `RecipesListComponent` by making a GET request to the `/api/recipes` service, as explained in *Chapter 3, Fetching Data as Streams*.

However, if we were handling a lazy loading mechanism, where we fetch just a few items (let's say 10) when the component loads initially, and then load more items as needed when the user clicks the **Next** or **Previous** buttons, as shown in the following screenshot, we'd need to adjust our logic:

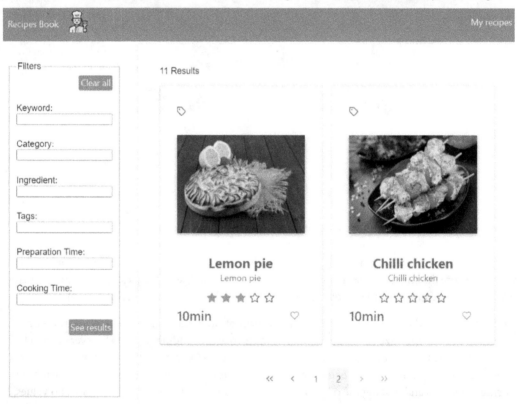

Figure 6.5 – Recipes list pagination

This would involve sending a GET HTTP request to fetch the data for the next page, using a URL structure such as `GET /api/recipes?page=1&limit=10`.

In such a scenario, `concatMap` is a very good option to issue a request for each emitted "next page" event, as follows:

```
recipes$ = this.pageNumberChange$.pipe(
    concatMap((pageNumber) =>
      this.http.get<Recipe[]>(`${BASE_PATH}/recipes`, {
        params: {
          page: pageNumber,
          limit: 10,
        },
      })
    )
  );
```

Here, pageNumberChange$ is a BehaviorSubject subject that emits the current page number whenever the user clicks on the **Next** or **Previous** buttons. concatMap then triggers subsequent HTTP GET requests, sequentially fetching the next page's list based on the current page number and size limit parameters. This sequential handling ensures data integrity and systematic pagination flow.

To summarize, concatMap is the ideal choice when you want to ensure that operations are processed sequentially, and that each inner Observable is processed *one at a time and in order*. However, when using concatMap, it's important to ensure that the inner Observable completes since concatMap waits for the completion of each inner Observable before subscribing to the next one in the sequence. If an inner Observable never completes, concatMap will also never subscribe to the subsequent Observables in the sequence. This can lead to blocking subsequent emissions and potential memory leaks or performance issues if there is a buildup of pending Observables. So, never use concatMap for endless streams.

It's important to note that not all higher-order mapping operators follow the concat strategy. There are other higher-order mapping operators, such as switch, merge, and exhaust, that offer different strategies and are useful in many situations. We'll break down those operators and their respective strategies in the following sections.

### The switchMap operator

switchMap is a combination of the switch and transformation (or mapping) strategies:

*switchMap = switch(switch) + map (higher-order mapping)*

Let's look at the marble diagram of the `switch` operator to understand the switch strategy:

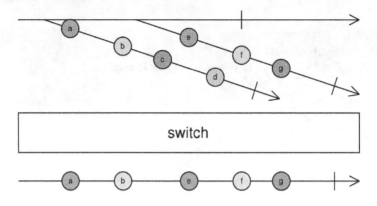

Figure 6.6 – The switch operator – marble diagram

Let's break down what's happening here (you are not used to seeing those diagonal lines, I know!):

- The top line is the higher-order Observable. The higher-order Observable emits the first inner Observable (which has the values **a**, **b**, **c**, and **d**). The switch operator subscribes to it under the hood.

- The first inner Observable emits the values **a** and **b**, and they get reflected automatically to the resulting Observable.

- Then, the higher-order Observable emits the second inner Observable (which has the values **e**, **f**, and **g**).

- The switch will unsubscribe from the first inner Observable (**a-b-c-d**) and subscribe to the second inner Observable (**e-f-g**); that's why the values **e**, **f**, and **g** get reflected right after **a** and **b**. As you may have noticed, in switching, if a new Observable starts emitting values, then the switch will subscribe to the new Observable and unsubscribe from the previous one.

So, the `switchMap` operator is a higher-order mapping operator that unsubscribes from any prior inner Observable and switches to any new inner Observable. It is useful when you want to cancel an operation when a new one is triggered. In other words, `switchMap` only focuses on the most recent data, ensuring that only the latest updates are processed while canceling any ongoing operations triggered by previous data.

Think of `switchMap` as changing TV channels: each time you press a button, you switch to a different channel, ignoring whatever was playing before. Similarly, `switchMap` lets you dynamically switch to a new Observable stream whenever the source emits, discarding any ongoing processing from previous emissions.

## Using switchMap for autosaving

Going back to our autosave reactive implementation in the Recipe app, if you're interested in saving the most recent form value and want to cancel any ongoing save operation if a new one is initiated before the current one finishes, then `switchMap` is the operator to use (instead of `concatMap`):

```
valueChanges$ = this.recipeForm.valueChanges.pipe(
    switchMap(formValue =>
      this.service.saveRecipe(<Recipe>formValue)),
    catchError(errors => of(errors)),
    tap(result => this.saveSuccess(result))
  );
```

As mentioned previously, I introduced a slight delay in the backend save service to illustrate how ongoing requests are handled when subsequent requests are made concurrently. So, when inspecting the network console, you'll notice the following:

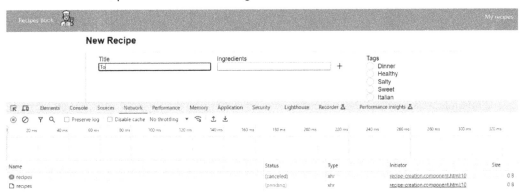

Figure 6.7 – The switchMap network requests

Here, we have two requests: one is marked as **canceled**, while the other is **pending**. The pending request is the most recent one, indicating that it was initiated while the previous request was still in progress, resulting in the cancellation of the prior request. And this is the behavior we aim for.

In this example, we're making an HTTP POST request, but we're only interested in the success or failure status and don't need any other data from the response. However, if you're expecting a response from the POST request to update the UI or perform tasks, keep in mind that *only the response from the latest request will be propagated*. In such cases, it's better to use `concatMap` instead.

## Using switchMap for autocompletion

Now, let's explore another practical scenario of using switchMap: autocompleting suggestions. This is a very common feature in web applications. In our recipe app, we'll implement an autocomplete dropdown for the tags field in RecipeCreationComponent. Currently, this field is displayed as a radio button with static values retrieved from the constant tags defined in src/app/core/ model/tags.ts. However, we'll transform it into a user-friendly autocomplete dropdown that dynamically fetches tag suggestions based on the user's input.

Whenever a user types a query, we'll retrieve the corresponding tags from a backend service. We have already implemented a service for that in our ready-to-use backend – that is, recipes-book-api. This service has an endpoint called /api/tags, accepts some criteria (the user's typed input) as a query parameter, and returns a list of tags that match the provided criteria. The code is available in this book's GitHub repository.

Let's delve into the implementation details. First and foremost, let's prepare our streams. How many streams do we have? We have two:

- A stream, named searchTerms, that emits the user's input, represented by a BehaviorSubject subject that's initialized with an empty string:

  ```
  private searchTerms = new BehaviorSubject<string>('');
  ```

  We'll update this stream whenever the user's input changes by using an update method:

  ```
  updateSearchTerm(searchTerm: string) {
    this.searchTerms.next(searchTerm);
  }
  ```

  The searchTerms stream and the updateSearchTerm method will be available in RecipeCreationComponent.

- A stream, named getTags$, that emits the fetched tags matching the user's input. We'll define this stream in RecipesService, as follows:

  ```
  getTags$: (term: string) => Observable<Tag[]> =
  (term: string) => {
    return this.http.get<Tag[]>(`${BASE_PATH}/tags`,
      { params: { criteria: term } });
  };
  ```

  getTags$ represents a function that takes a string parameter, term, and returns an Observable of the Tag[] type that issues an HTTP GET request to retrieve an array of Tag objects matching the provided search term.

  We defined the Tag type in \src\app\core\model\tags.ts.

Now, it's time to use `switchMap`, which will map each value emitted by the `searchTerms` stream to the `getTags$` Observable:

```
tagValues$ = this.searchTerms.pipe(
distinctUntilChanged(), // ignore if next search term is
                             same as previous
switchMap((term: string) => this.service.getTags$(term))
// switch to new Observable each time
  );
```

We will define `tagValues$` in `RecipeCreationComponent`. So, overall, `tagValues$` issues a search request for every unique user's input and ensures only the latest search results are displayed, discarding previous ones.

Finally, we will update the `RecipeCreationComponent` HTML template to modify how tags are displayed from a radio button to an autocomplete dropdown:

```
<div class="col-3">
    <label for="Tags">Tags</label>
    @if (tagValues$ | async; as tags) {
        <p-autoComplete formControlName="tags"
        [suggestions]="tags"
        (completeMethod)=
        "updateSearchTerm($event.query)"
        field="name"></p-autoComplete>
    }
</div>
```

Here, we subscribed to `tagValues$` using the async pipe and stored the emitted value in a `tags` array. Then, we used the PrimeNG autocomplete component to provide suggestions as the user types in the `"Tags"` input field. The autocomplete component binds to a form control named `"tags"`, receives suggestions from the `tags` array, and triggers the `updateSearchTerm` method with the user's query when the user starts typing. `concatMap` will issue a GET request for every unique user's input and ensure only the latest search results are displayed, canceling previous requests.

And that's it! Here's an illustration of the implemented behavior. Here, when we enter **B** in the search field, we receive **Breakfast** a suggestion:

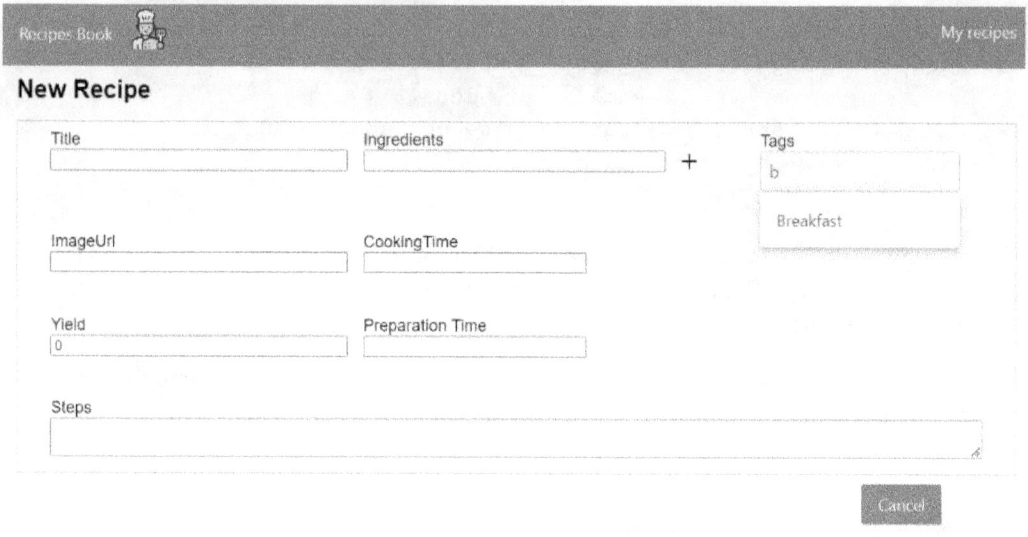

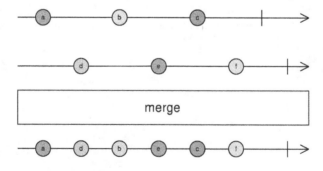

Figure 6.8 – The autocompletion suggestion

Now, let's move on to another operator, mergeMap.

### The mergeMap operator

mergeMap is a combination of the merge and transformation (or mapping) strategies:

*mergeMap = merge(merge) + map (higher-order mapping)*

Now that you understand the concepts of higher-order mapping, let's understand the merging strategy by looking at the following marble diagram, which considers the merge operator:

Figure 6.9 – The merge operator – marble diagram

Unlike `concat`, `merge` will not wait for an Observable to complete before subscribing to the next Observable. It subscribes to every inner Observable at the same time and then outputs the values to the combined result. As described in this marble diagram, the values of the input Observables are reflected in the output immediately. The result will only be completed once all the merged Observables are completed.

`mergeMap` is a higher-order mapping operator that processes each inner Observable in parallel. It is like multitasking in the kitchen, where you're simultaneously handling different cooking tasks, such as chopping, boiling, and mixing, all at once. However, you should only use `mergeMap` if the resulting order doesn't matter because these requests may be processed out of order.

Suppose we aim to retrieve a list of recipes that match specific tags. For every tag, we want to initiate an HTTP request to fetch the recipes corresponding to that tag. The order of tag requests is not important; all requests should be executed concurrently. Here's the code:

```
selectedTags$ = from(['Salty', 'Sweet', 'Healthy']);

recipesByTag$ = this.selectedTags$.pipe(
  mergeMap(tag =>
    this.getRecipesByTag(tag)),mergeAll(),toArray());
getRecipesByTag(name: string): Observable<Recipe[]> {
  return this.http.get<Recipe[]>(
    `${BASE_PATH}/recipesByTags`, { params: { tagName:
      name } });
}
```

Here, we created an Observable named `selectedTags$` from a static array of tags. `selectedTags$` emits the tags (array elements) one by one. Whenever a tag is emitted by `selectedTags$`, an HTTP request is issued by `this.getRecipesByTag(tagName)` to fetch the corresponding recipes. The `mergeMap` operator is used to concurrently handle multiple tag requests. When another tag is emitted while the previous request is still in progress, the new request is executed concurrently.

We used the `mergeAll` operator to flatten the results that were obtained from the different inner Observables into a single Observable stream. This ensures that the recipes emitted by each inner Observable are merged into a cohesive stream of recipes.

Finally, the `toArray` operator is used to convert all emitted recipes into a single array, making it convenient for a list display in the UI.

Again, I previously added a delay to the backend service that returns the recipes by tag. Upon opening the console, we'll find that all the requests were run concurrently, even if there are pending ones:

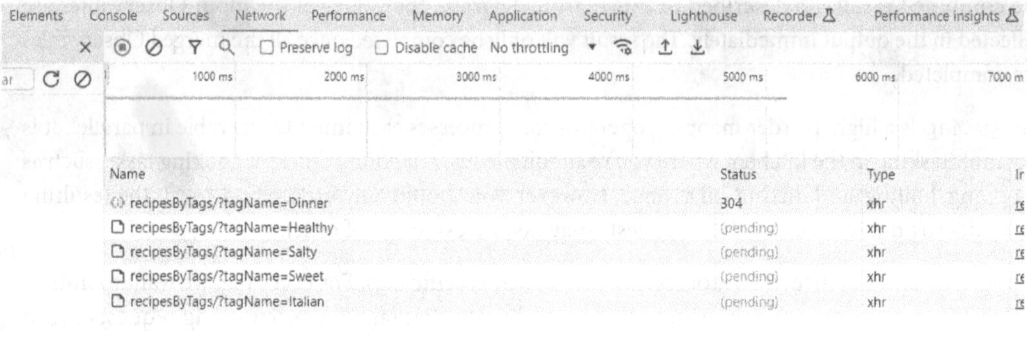

Figure 6.10 – mergeMap's execution

We can also use `mergeMap` to fetch data from multiple sources in parallel and combine the results. Imagine that we have multiple Review sources for our recipes, and we want to collect them (in this case, the order here is not important). Here's the code:

```
getRecipesReviews(recipeId: number): Observable<Review[]>
{
  return from([`${BASE_PATH}/source1/reviews`,
  `${BASE_PATH}/source2/reviews`])
    .pipe(
      mergeMap((endpoint) => this.http.get<Review[]>(
        endpoint, { params: { recipeId: recipeId } })));
}
```

Here, `getRecipesReviews(recipeId: number)` is a method that fetches reviews for a recipe identified by `recipeId` from two different sources (`source1` and `source2`) by issuing parallel HTTP GET requests to each source.

The `from([${BASE_PATH}/source1/reviews, ${BASE_PATH}/source2/reviews])` line creates an Observable from an array containing two different endpoints to fetch reviews. The `from` operator emits each item of the array as a separate value in the Observable sequence.

Then, `mergeMap` is used to raise a GET request to fetch reviews for the specified `recipeId` for each endpoint emitted by the `from` operator, ensuring that both requests are made concurrently.

With that, let's move on to the last operator that we will discuss, `exhaustMap`.

## The exhaustMap operator

exhaustMap is a combination of the exhaust and transformation (or mapping) strategies:

*exhaustMap = exhaust(exhaust) + map ()higher-order mapping*

Let's look at this marble diagram to understand the exhaust strategy:

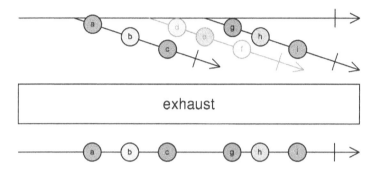

Figure 6.11 – The exhaust operator – marble diagram

The top line is a higher-order Observable that produces three inner Observables over time. When the first inner Observable (**a-b-c**) is emitted, exhaust will subscribe to it so that the values **a-b** get reflected in the output Observable. Then, the second inner Observable comes and gets ignored by the exhaust operator; it will not be subscribed to (this is the key part of exhaust).

Only when the first inner Observable completes will exhaust subscribe to new Observables. So, **c** will be reflected, and the first inner Observable will complete. exhaust is ready now to treat other Observables. At that point, the third inner Observable comes. The switch will subscribe to and the values of the third inner Observable, **g-h-i**, get reflected in the output.

So, exhaustMap waits for the completion of the current inner Observable before allowing the next Observable to emit values. Once the inner Observable completes, exhaustMap subscribes to the next Observable in the sequence. It ensures that only one inner Observable is active at a time, ignoring any new Observables that are emitted while the current one is still ongoing. This is particularly useful in scenarios where you want to ignore new events until a previous operation has finished, such as handling user clicks on a button, where subsequent clicks are ignored until the current operation completes.

It's similar to how you might handle tasks when you're busy with something important. If someone tries to get your attention with a new task, you might say, "I'm busy right now, please don't disturb me until I finish what I'm doing." exhaustMap operates similarly, ensuring that ongoing tasks are completed before considering new ones.

Let's consider a scenario in our recipe app where users can edit recipe details and save their changes through a **Save** button. You want to prevent multiple save requests from being sent if a user rapidly clicks the **Save** button multiple times. `exhaustMap` comes in handy here by ignoring subsequent save requests until the current save operation completes. Here's the implementation:

```
private saveClick = new Subject<Boolean>();
private saveRecipe$ =
  this.service.saveRecipe(<Recipe>this.recipeForm.value);
saveClick$ = this.saveClick.pipe(exhaustMap(() =>
  this.service.saveRecipe(<Recipe>this.recipeForm.value))
);
saveRecipe() {
  this.saveClick.next(true);
}
```

In the component responsible for editing the recipe, we defined the following:

- A private `saveClick` subject to track the **Save** button being clicked. `saveClick` represents our first stream.

- A private `saveRecipe$` Observable that issues an HTTP save request to save the recipe.

- `saveClick$`, our second stream that listens to `saveClick` emissions and uses `exhaustMap` operator to issue a new save request only after the previous one is completed.

- A `saveRecipe` method. This will be invoked when the **Save** button is clicked and will send `true` to the `saveClick` subject.

Finally, we must subscribe to `saveClick$` using the async pipe in the HTML template and add the click handler to the **Save** button to invoke the `saveRecipe` method, as follows:

```
<ng-container *ngIf="saveClick$ | async"></ng-container>
<p-button class="recipe-button" (click)="saveRecipe()"
label="Save"></p-button>
```

This ensures that only one save request is processed at a time, preventing duplicate entries or data corruption. You can test this code in `RecipeCreationComponent`.

You can also use `exhaustMap` in drag and drop features to ensure that actions are only processed when a user's dragging action has been completed, preventing multiple actions from being triggered simultaneously.

### *Wrapping up the operators*

Let's summarize all the operators that were mentioned in this chapter:

- If the order is important and you need to process operations in sequence while waiting for completion, then `concatMap` is the right choice
- If the order is not important and you need to process operations in parallel to enhance performance, `mergeMap` is the best operator
- If you need to put a cancellation logic to release resources and take always the most recent information, then `switchMap` is the way to go
- To ignore new Observables while the current one is still ongoing, use `exhaustMap`

All you have to do is pick the right operator based on your specific use case.

## Summary

In this chapter, we began by illustrating the traditional, imperative approach to implementing the autosave feature in our recipe app. However, we soon encountered limitations with this method. We highlighted these problems before exploring a more reactive pattern to address these challenges.

Then, we delved into higher-order Observables and higher-order mapping operators, learning how the `concatMap` operator works and how it can help us implement the autosave requirement in the Recipe app reactively.

Additionally, we expanded our exploration to include other strategies, namely the `merge`, `switch`, and `exhaust` higher-order mapping operators. We explained their functionality by using practical examples and use cases to gain a deeper understanding of these concepts.

In the next chapter, we will explore another useful reactive pattern that allows you to share data between your components. As usual, we will demystify the concepts and then learn the reactive way to do it.

# 7

# Sharing Data between Angular Components

Sharing data between components is a very common use case in web applications. Angular provides many approaches for communicating between parent and child components, such as the popular @Input() and @Output() decorator patterns. The @Input() decorator allows parent components to provide data to their child components, while the @Output() decorator allows the child component to send data to a parent component. That's great, but when data needs to be shared between components that are either deeply nested or not immediately connected, those kinds of techniques become less efficient and difficult to maintain.

So, what's the best way to share data between sibling components? This is the heart of this chapter. We will start by explaining the sharing data requirement, before walking through the different steps to implement the reactive pattern for sharing data between sibling components in our app. Finally, we will introduce Angular's new Deferrable Views feature to maximize our app's performance.

In this chapter, we're going to cover the following main topics:

- Defining the sharing data requirement
- Exploring the reactive pattern to share data
- Leveraging Deferrable Views in Angular 17

## Technical requirements

This chapter assumes that you have a basic understanding of RxJS.

The source code of this chapter is available at `https://github.com/PacktPublishing/Reactive-Patterns-with-RxJS-and-Angular-Signals-Second-Edition/tree/main/Chap07`.

## Defining the sharing data requirement

Let's assume that we have four components – **C1**, **C2**, **C3**, and **C4** – that do not have any relationship with each other, and there is information – **DATA** – shared between those components:

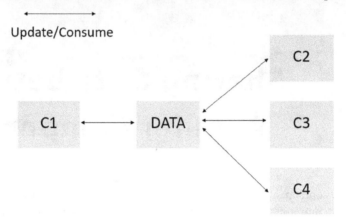

Figure 7.1 – Shared data between components

The components can update and consume **DATA** at the same time. But at any time during the process, the components should be able to access the last value of **DATA**.

Now, let's make that explanation clearer with a more concrete example.

In our recipe application, when the user clicks on one recipe, it gets selected, but we want all components to have access to the last selected recipe by the user. In that case, the selected recipe represents our shared data.

One of the components that will need access to the selected recipe is the `RecipeDetailsComponent` component, as it will display the details of the selected recipe.

Without further ado, in the next section, let's see how we can make this data available to everyone in a reactive way.

## Exploring the reactive pattern to share data

Angular services are powerful and efficient for creating common references to share both data and business logic between components. We will combine Angular services with Observables – more specifically, `BehaviorSubject` instances – to create stateful, reactive services that will allow us to synchronize the state efficiently across an entire application. So, in the following subsections, let's explain the steps to implement a reactive pattern to share data between unrelated or sibling components.

## Step 1 – Creating a shared service

First, we will create an Angular service called `SharedDataService` using the Angular CLI, as usual under the `src/app/core/services` folder:

```
ng g s SharedData
```

> **Note**
>
> Here, we named the service `SharedDataService` for demonstration purposes. While it's true that we already have a service named `RecipesService` that could have accommodated the shared data, the purpose of this chapter is to underscore the broader concept of data sharing. Therefore, we chose a more generic term. However, in your own application, it's recommended to use specific and descriptive names such as `RecipesService` or another name that accurately reflects the role and domain of the service. A name that accurately reflects the purpose of your service is crucial for clarity and maintainability, especially in a framework such as Angular where conventions can guide developers.

Then, in the `SharedDataService` class, we need to create the following:

- A private `BehaviorSubject` instance called `selectedRecipeSubject` that emits the value of the currently selected recipe, which represents the data to be shared:

  ```
  private selectedRecipeSubject = new BehaviorSubject<Recipe |
  undefined>(undefined);
  ```

  Here, `selectedRecipeSubject` has `Recipe` as the type and `undefined` as the initial value since initially, we don't have any selected value.

  Also, `selectedRecipeSubject` is declared as `private` to ensure that it's only accessible within `SharedDataService`, where it's defined, protecting it from external manipulation. Otherwise, any external process could have access to the property and consequently call the `next` method and change the emissions, which is dangerous. This encapsulation is important for maintaining control over the state and preventing unintended changes.

- A public Observable, named `selectedRecipe$`, extracted from `selectedRecipeSubject` to handle data as an Observable:

  ```
  selectedRecipe$ = this.selectedRecipeSubject.asObservable();
  ```

  Here, we used the `asObservable()` method available in the `Subject` type to derive a read-only Observable from `selectedRecipeSubject`. This ensures that the emissions of `selectedRecipeSubject` are only consumed in read-only mode, preventing external processes from altering the `selectedRecipeSubject` value.

- A method called `updateSelectedRecipe` that will update the shared data, which is the selected recipe:

```
updateSelectedRecipe(recipe: Recipe) {
  this.selectedRecipeSubject.next(recipe);
}
```

This method only calls `next` on `selectedRecipeSubject` to notify all subscribers of the last selected recipe passed as a parameter. The process that updates the selected recipe will call this method, which we will discuss in the next step.

This is what the service looks like after putting all the pieces together:

```
import { Injectable } from '@angular/core';
import { BehaviorSubject } from 'rxjs';
import { Recipe } from '../model/recipe.model';

@Injectable({
  providedIn: 'root'
})
export class SharedDataService {

  private selectedRecipeSubject = new
    BehaviorSubject<Recipe | undefined>(undefined);
  selectedRecipe$ =
    this.selectedRecipeSubject.asObservable();

  updateSelectedRecipe(recipe: Recipe) {
    this.selectedRecipeSubject.next(recipe);
  }
}
```

Now that we have prepared the groundwork by creating our shared data service and defined the behavior subject that will hold the shared data, let's see how we can update the shared data in the next section.

## Step 2 – Updating the last selected recipe

We should update the shared `selectedRecipe` instance when the user clicks on one of the recipe cards in the `RecipesListComponent` component. As a reminder, here are the recipe cards in our `Recipe` app:

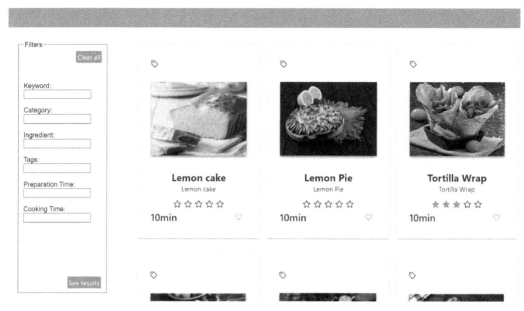

Figure 7.2 – List of recipes

In order to update the shared `selectedRecipe` instance when the user clicks on the card, we need to incorporate the `(click)` event output in the `RecipesListComponent` HTML template, which triggers the execution of the `editRecipe(recipe)` method. This is the HTML code required:

```
@if (filteredRecipes$ | async; as recipes) {
<div class="card">
    <p-dataView #dv [value]="recipes" [paginator]="true"
    [rows]="9" filterBy="name" layout="grid">
        <ng-template let-recipes pTemplate="gridItem">
            <div class="grid grid-nogutter">
                @for (recipe of recipes; track recipe) {
                <div class="col-12" style="cursor:
                    pointer;" (click)="editRecipe(recipe)"
                        class="recipe-grid-item card">
// extra code here
</div>
} @empty {
<div>There are no recipes</div>
}
            </div>
        </ng-template>
    </p-dataView>
</div>
```

```
} @else {
<div>There are no recipes</div>
}
```

Here, (click) event binding is applied to each card, ensuring that when clicked, the editRecipe(recipe) method is invoked to update the selectedRecipe instance.

In RecipesListComponent, we implement the editRecipe method as follows:

```
editRecipe(recipe: Recipe) {
  this.sharedService.updateSelectedRecipe(recipe);
  this.router.navigate(['/recipes/details']);
}
```

The editRecipe method takes the selected recipe as the input and performs two actions:

- It notifies selectedRecipeSubject that the value of selectedRecipe has changed by calling the updateSelectedRecipe(recipe:Recipe) method, available in SharedDataService. So, we should inject the SharedDataService service in RecipesListComponent as follows:

  ```
  import { SharedDataService } from '../core/services/shared-data.
  service';
  export class RecipesListComponent implements OnInit {
    constructor(private sharedService:
    SharedDataService) {}
  }
  ```

- It displays the details of the recipe by routing to RecipeDetailsComponent, the standalone component responsible for rendering and displaying the details of a recipe. We've added a route configuration in the app-routing-module.ts file as follows:

  ```
  import { RecipeDetailsComponent } from
  './recipe-details/recipe-details.component';
  const routes: Routes = [
    { path: 'recipes/details',
    component: RecipeDetailsComponent},
  ];
  ```

At this point, we have put in place the mechanism that updates the value of the shared data. Now, all that is left is to listen to the shared data and consume it.

## Step 3 – Consuming the last selected recipe

In the `RecipeDetails` component, we need to consume the last selected recipe in order to display its details. So, again, we need to inject `SharedDataService` and define the `selectedRecipe$` Observable – which will emit the last selected recipe – as follows:

```
import { SharedDataService } from '../core/services/shared-data.
service';
export class RecipeDetailsComponent {
  constructor(private sharedService: SharedDataService) { }
  selectedRecipe$ = this.sharedService.selectedRecipe$;
}
```

Then, we will subscribe to the `selectedRecipe$` Observable using the `async` pipe in the `RecipeDetailsComponent` HTML template in order to display the selected recipe's details, as follows :

```
@if (selectedRecipe$|async;as recipe) {
<div>
    <span> {{recipe.title}} </span>
    <span> {{recipe.steps}} </span>
    <span> {{recipe.ingredients}} </span>
</div>
}
```

And that's it – this is how you can share data between unrelated components throughout the application!

Now, we can use the latest value of the recipe everywhere – we only have to inject `SharedDataService` into the component that needs the shared data and subscribe to the public Observable that emits the read-only value. For example, we can add this code in `HeaderComponent` to display the title of the last selected recipe in the application's header:

```
@if(selectedRecipe$|async; as recipe) {
    <div>
        <span> {{recipe.title}} </span>
    </div>
}
```

If we change the shared value in this component, all other components that listen to the shared data will get notified to update their processes.

> **Note**
>
> We used this pattern in *Chapter 5*, *Combining Streams*, to share the value of the filter in `RecipesFilterComponent` with `RecipesListComponent` instances, and then we combined the streams to display the filtered results.

## Wrapping up the data-sharing reactive pattern

To summarize everything, here's a wrap-up of the steps:

- Begin by creating an Angular service that will be shared across components. Within this service, define a private `BehaviorSubject` instance that will emit the shared value to its subscribers, remembering to specify the type of data emitted by `BehaviorSubject` and initialize it with the initial value of the shared data.

  It's important to note that we use `BehaviorSubject` for two primary reasons:

  - It allows us to broadcast shared data to multiple observers.

  - It stores the latest value emitted to its observers, and any new subscriber immediately receives the last emitted value upon subscription.

- Next, define a public Observable within the shared service to hold the read-only shared value.

- Implement an `update` method within the shared service to update the shared value by calling the `next` method of the `Subject` type to emit the updated value to subscribers.

- Inject the shared service in the component responsible for updating the value of the shared data and call the `update` method implemented in the service.

- Inject the shared service in the component that consumes the value of the shared data and subscribe to the exposed Observable in the service.

This reactive sharing data pattern has many benefits:

- It improves the sharing of data between unrelated components.

- It manages mutability risk. In fact, as we only expose the read-only extracted Observable to other consumers and keep `BehaviorSubject` private, we prevent shared data from being modified by subscribers, which can lead to data corruption and unexpected behavior.

- It makes communication between components easier as you will only have to inject the shared service where you need it and just take care of updating the data.

As far as I'm concerned, this is the simplest way to share data between unrelated components in Angular and manage the application state. This works perfectly in many situations, but for big applications where there are a lot of user interactions and multiple data sources, managing states in services can become complicated.

In those cases, we can use a state management library to manage the state of our application. There are many great state management libraries out there to manage states in Angular, all with one thing in common – they are built on top of RxJS Observables, and the state is stored in `BehaviorSubject`. The most popular state management library is NgRx, which you can find out more about here: `https://ngrx.io/guide/store`.

Data-sharing mechanisms facilitate communication between different components and improve the user experience, as well as the responsiveness of your application. Before ending this chapter, I want to shed light on a new feature introduced in Angular 17, Deferrable Views, which can complement data sharing and contribute to creating more responsive and efficient applications. Let's see this in action in the next section.

## Leveraging Deferrable Views in Angular 17

**Deferrable Views** allow you to declaratively mark parts of your templates as non-essential for immediate rendering. It is kind like delaying the rendering of certain parts of a page to improve the perceived performance of your application, as well as optimize the initial bundle size and loading times.

There are a number of real-world scenarios where defer rendering can help to achieve faster load times such as e-commerce product pages – in this example, you can initially display the essential product details and then lazy load additional content such as reviews when the user clicks on a **Read more** button or scrolls down the page.

Let's quickly see how this works. To lazy load a component, you need to use a Standalone component, otherwise deferring won't work. Then you want to wrap up the Standalone component in a `@defer` block, like so:

```
@defer {
  <delayed-component />
}
```

You can also define conditions for when exactly the deferred component should load. You can do this by using **triggers**, which specify events or situations that initiate loading:

```
@defer(on viewport) {
  <delayed-component />
}
```

Here, the `on viewport` trigger is used to display the `delayed-component` when it enters the viewport area of the user's browser window.

Besides `on viewport`, there are other triggers that can be used such as `on hover`, which only initiates the content to load when the user's mouse hovers over the delayed content. You can find a full list of available triggers here: `https://angular.dev/guide/defer#triggers`.

Furthermore, the `@defer` block has some important sub-blocks. For example, you can display alternative content before the deferred content loads using the `@placeholder` sub-block, like so:

```
@defer(on viewport) {
    <delayed-component />
}
@placeholder {
    <div>Placeholder text here</div>
}
```

In addition to @placeholder, @defer also offers two other sub-blocks – @loading and @error:

- The @loading block is similar to the @placeholder block, but it specifically shows content like a loading message while the actual content is being prepared.

- The @error block is displayed if there is an error while fetching or processing the deferred content. This allows you to provide a user-friendly error message, or alternative content, in case something goes wrong.

Now let's look at how we can utilize defer rending in our Recipe app. Given that the images for each recipe have a high resolution, let's delay the rendering of the images in RecipesListComponent HTML template so that they are only shown when the user hovers over the image:

```
@defer (on hover) {
    <img class="recipe-image" [src]="'assets/recipes/'+
    recipe.imageUrl" [alt]="recipe.title" />
}
@placeholder {
    <div>Hover to load the image</div>
}
```

As you can see, we surrounded the code block displaying the image with a @defer block, and used the on hover trigger. Then we used the @placeholder block to specify some text that should be displayed while the deferred content is not yet loaded. In this case, inside the <div> element, we added the text, **Hover to load the image**.

For more details about the Deferrable Views feature, refer to https://angular.dev/guide/image-optimization.

## Summary

In this chapter, we explained the motivation behind sharing data between components and learned how to implement it in a reactive way. First, we learned how we can use BehaviorSubject combined with Angular services to share data between unrelated components and manage our application state. Then, we highlighted the advantages of the sharing data reactive pattern. Finally, we explored Angular's new Deferrable Views feature.

The features covered in this chapter will help you implement a good architecture for your web application, making it more reactive and performant, improving load times, and reducing the cost of maintainability.

Now, get ready for an exciting journey, because in the next chapter, we delve into a brand-new feature called Angular Signals! We'll cover some reactive patterns using Signals and even integrate them into what we've learned so far.

# Part 3:
# The Power of
# Angular Signals

Dive into the exciting world of Angular Signals!

In this section, you will discover the core functionalities and advantages of Angular Signals, as well as unlock the potential of reactivity by leveraging Angular Signals and RxJS together. We will also go through the latest Angular Signals improvements.

This part includes the following chapter:

- *Chapter 8, Mastering Reactivity with Angular Signals*

# 8

# Mastering Reactivity with Angular Signals

Modern web applications thrive on reactivity, where data changes automatically, thereby triggering updates in the UI. Angular Signals, introduced in version 17, streamlines this process by offering a powerful and concise way to manage reactive data within your Angular applications.

This chapter delves into the core concepts, API functionalities, advantages, and relationship of Signals with RxJS. We will also see how we can improve the reactivity of our Recipe app further using Angular Signals.

So, in this chapter, we're going to cover the following main topics:

- Understanding the motivation behind Signals
- Unveiling the Signal API
- Unlocking the Power of RxJS and Angular Signals
- Integrating Signals into our recipe app
- Reactive data binding with Signals

## Technical requirements

The source code for this chapter is available at `https://github.com/PacktPublishing/Reactive-Patterns-with-RxJS-and-Angular-Signals-Second-Edition/tree/main/Chap08` (this only includes the code related to the recipe app).

# Understanding the motivation behind Signals

The main goal behind the Angular team's introduction of Signals is to add more fine-grained reactivity to the framework. This new Signal-based reactive system marks a significant leap forward in the framework's ability to handle dynamic data and user interactions. It offers a fresh approach to detecting and triggering changes within the framework, replacing the traditional approach that relies on Zone.js.

## The traditional Zone.js approach

Angular's traditional change detection mechanism assumes that any event handler can potentially change any bound data to the template. That's why, whenever an event happens in your Angular application, the framework scans all components and their data bindings for any potential changes. This can be a bit heavy-handed, especially for complex applications. For this reason, a more optimized mode OnPush change detection was introduced. This mode leverages the concepts of immutability and Observables, allowing Angular to significantly reduce the number of components it needs to check for updates. This was explored in *Chapter 3, Fetching Data as Streams.*

Whether you use the default change detection or the more optimized OnPush mode, Angular still needs to stay informed when event handlers have finished running. This presents a challenge because the browser – not Angular itself – triggers these event handlers. This is where Zone.js steps in, essentially acting as a bridge. Zone.js can detect when an event handler has run, telling Angular, "Hey, there's a new event; you can take care of any necessary updates now."

While this approach has worked well in the past, it still comes with a few downsides: when changes are made, the whole component tree and all the expressions on every component are always checked. There is no way for Angular to directly identify changed components or to just update the changed parts of a component. That is why Angular cannot make any assumptions about what happened and needs to check everything!

## The new Signals approach

With Signals, Angular can easily detect when any part of the application data changes and update any dependencies automatically. Signals enable efficient change detection, smarter re-rendering when data changes, and facilitate fine-grained updates to the DOM, reducing the runtime required for Angular to check all components, even if their consumed data remains unchanged. Ultimately, this can eliminate the need for Zone.js in one of the future versions:

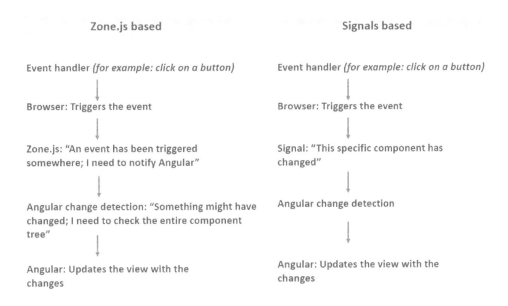

**Zone.js based**

Event handler *(for example: click on a button)*

↓

Browser: Triggers the event

↓

Zone.js: "An event has been triggered somewhere; I need to notify Angular"

↓

Angular change detection: "Something might have changed; I need to check the entire component tree"

↓

Angular: Updates the view with the changes

**Signals based**

Event handler *(for example: click on a button)*

↓

Browser: Triggers the event

↓

Signal: "This specific component has changed"

↓

Angular change detection

↓

Angular: Updates the view with the changes

Figure 8.1: Comparing the various change detection approaches

Apart from improving change detection, there are other advantages of using Signals:

- It provides a more intuitive and declarative way to manage reactive data

- The syntax aligns more closely with JavaScript, making code easier to read, understand, and maintain.

- The compiler performs better type narrowing for improved type safety within your reactive code.

As we progress through this chapter, you'll gain a clearer understanding of Signals. Eager to discover more? Let's continue.

# Unveiling the Signal API

In this section, we'll delve into the world of Signals, covering exactly what they are, how they work, and the revolutionary changes they bring to Angular. So, without further ado, let's discover what a Signal is.

## Defining Signals

A **Signal** is a reactive entity within Angular that encapsulates a value (serving as a container for a value) and automatically notifies consumers whenever that value changes. You can think of Angular's Signals as a combination of a data value and a change notification mechanism, offering a streamlined approach to tracking changes and seamlessly updating the user interface in response to those changes.

While the concept of Signals is not novel and has existed in various forms across different frameworks for many years, their integration into Angular provides developers with a familiar yet powerful tool for managing reactive behavior within their applications. They act as wrappers around values, allowing you to efficiently track changes and react accordingly.

## Creating Signals using the constructor function

We can create a Signal using the `signal` constructor function available in the `@angular/core` package.

An initial value is always required as a Signal must always have a value. Signals can hold a wide range of values, including simple primitives, such as strings and numbers, as well as more complex data structures, such as arrays and objects.

Plus, the `signal` function provides type flexibility. You can either explicitly define the type of the Signal's value or leverage type inference based on the initial value. For example, the following code creates and initializes a Signal with a value of John Doe:

```
import { signal } from '@angular/core';
const name=signal('John Doe');
```

In this example, we didn't define a type for our Signal's value. If you don't explicitly specify a type, the `signal` function can infer the type based on the initial value you provide. So, here, the type can be inferred from the initial value of John Doe as `string`.

But what if you want to be extra clear about the type? That's perfectly possible! Here's how you would explicitly define the type:

```
name = signal<string>('John Doe');
```

As you can see, we added `<string>` after `signal` to explicitly state that the Signal will hold string values. While type inference works well in many cases, explicitly defining types can improve code readability and maintainability, especially for larger projects.

Now, let's look at an example of a Signal holding an array. Imagine that you want to define a Signal that represents an array of currencies:

```
currencies=signal(['EURO', 'DOLLAR', 'Japanese yen', 'Sterling'])
```

Here, the initial value is an array of strings, so the type of the Signal will be inferred as an array of strings, `string[]`.

Now, thinking of our Recipe app, the `favouriteRecipe` Signal holds a `Recipe` object and is of the `Recipe` type:

```
favouriteRecipe=signal<Recipe>({
   id: 1,
   title: "Lemon cake",
   prepTime: 10,
   cookingTime: 35,
   yield: 10,
   imageUrl: "lemon-cake.jpg"
})
```

Here, we explicitly defined the specific type of `Recipe`. If you're working with a specific type throughout your code, explicit type definition provides clarity and prevents potential type mismatches. However, when the type is different from the initial value or may vary, you can avoid explicitly writing it, making your code cleaner and more concise.

Once we have a Signal, we often want to read it and retrieve its value. But how can we do that? We'll find out in the next section.

## Reading Signals

One way to read a Signal's value is by using the Signal's getter. Here's an example that reads the value of the previously created Signals and logs it in the console:

```
console.log(this.name());
console.log(this.favouriteRecipe());
//console output
John Doe
{"id": 1,"title": "Lemon cake","prepTime": 10, "cookingTime":
35,"yield": 10,"imageUrl": "lemon- cake.jpg" }
}
```

You can use this getter to read Signals in your Angular components, services, and directives.

You can also read Signals in your component template to display a value:

```
@for (currency of currencies(); track currency) {
<option>{{currency}}</option>
} @empty {
<div>There are no currencies</div>
}
<div>{{favouriteRecipe().title}}</div>
```

Reading a Signal in a template returns the current Signal's value and registers the Signal as a dependency of the template. If the Signal changes, the portion of the template is re-rendered.

Signals that are created using the signal creator function are *writable*. This means you can modify their values after their creation. We'll learn how to modify the value of a Signal in the next section.

## Modifying a writable Signal

Signals that are created using the `creation` function are of the `WritableSignal` type and offer an API specifically for updating their values. There are two primary methods for modifying the stored value of a writable Signal:

- Using the `set` method
- Using the `update` method

Let's look at them both.

### Using the set method

The `set` method lets you directly set a new value for the Signal. Here's an example:

```
name = signal('John Doe');
console.log(this.name());
this.name.set('Mary Jane');
console.log(this.name());
//console output
John Doe
Mary Jane
```

Here, we used the `set` method to update the Signal's value from John Doe to Mary Jane. This is a simple and effective way to assign a new value when you know how the value needs to be changed.

### Using the update method

The `update` method allows us to compute a new value from the previous one, like so:

```
name = signal('John Doe');
console.log(this.name());
this.name.update(value=>'Full Name: '+ value);
console.log(this.name());
//console output
Full Name: John Doe
```

In the `update` method, we appended `Full Name:` to the old Signal's value, `John Doe`. The Signal maintains a record of value changes over time. When the value changes, the Signal notifies subscribed components or logic, prompting necessary UI or data flow modifications. Every part of the Angular component that depends on the Signal will be automatically updated once the value changes.

So far, so good! Now that you're comfortable with the basics of Signals, what if you could create Signals that automatically react to changes in other Signals? In other words, what if you need Signals that depend on other Signals? Well, that's where computed Signals come into play!

## Computed Signals

Computed Signals deduce their value from other Signals, offering a declarative way to define relationships between Signals and ensure your data remains consistent. Let's focus on a simple example to understand the behavior:

```
import { signal, computed } from '@angular/core';

const firstName = signal('John');
const lastName = signal('Doe');
const fullName = computed(() => `${firstName()} ${lastName()}`);
```

In this code block, the `fullName` computed Signal derives its value from both the `firstName` and `lastName` Signals. The `computed` function simply appends the `firstName` and `lastName` values. So, `fullName` depends on the `firstName` and `lastName` Signals, which means that whenever either `firstName` or `lastName` changes, the `fullName` Signal automatically updates, reflecting the complete name.

Note that the computed Signal is lazy evaluated and cached. This means that the `computed` function doesn't execute to calculate its value until the first time you read the calculated Signal (`fullName`, in our case). The calculated value is then cached, and if you read `fullName` again, it will return the cached value without re-executing the calculation function. Then, if the value of `firstName` or `lastName` changes, Angular knows that the `fullName` cached value is no longer valid, and the next time you read `fullName`, its new value will be re-calculated. So, the calculation function will re-execute again.

> **Note**
> Unlike `WritableSignals`, computed Signals are read-only, so you can't change their values. Even trying to set a value will result in a compilation error.

Now that we've looked at computed Signals, which automatically react to changes in other Signals, what if you need to perform actions beyond simply updating data, such as making API calls or interacting with other components? This is where Signal effects step in!

## Signal effects

Signal effects are functions that execute in response to Signal changes. They provide us with a way to perform side effects, such as logging data, or manipulating the DOM to perform custom rendering or adding a custom behavior.

Let's look at an example. Here's some code in the HTML template:

```
<button (click)="update()">Update</button>
```

And here's some TypeScript code :

```
counter = signal(0);
constructor() {
  effect(() => {
    console.log('The updated value is', this.counter());
  });
}
update() {
  this.counter.update((current) => current + 1);
}
```

This code creates a counter that starts at 0. Then, clicking the created **Update** button increments the counter value by 1, and effect logs the updated value to the console.

Note that effects require an injection context to function properly, such as during the construction of a component or service. That's why we called it inside the constructor in the previous example. This means it needs to be called within a specific environment where Angular's dependency injection system is available.

But why? Well, because Signal effects might internally rely on other Angular services or functionalities that are managed by the dependency injection system. So, we should ensure that all necessary dependencies are properly injected and accessible for effects to work as intended. Running it outside this context could lead to errors because these dependencies wouldn't be available.

Having explored the core functionalities and concepts of Signals, you might be wondering how they compare to RxJS. Both offer mechanisms for managing data streams, so how do they differ? And can they work together? These are crucial questions we'll address in the next section.

# Unlocking the power of RxJS and Angular Signals

While Angular Signals serves as a lightweight wrapper for reactive data with a simplified API, RxJS offers a comprehensive library for handling asynchronous streams, thus remaining crucial for handling more complex reactive programming requirements.

Here's a concise comparison of Signals and RxJS Observables:

| Feature | Signals | Observables |
|---------|---------|-------------|
| Value representation | Hold a single value at a time. | Emit values over time. |
| Subscription | Subscription is implicit. | Require explicit subscriptions. |
| Updates capability | Updated by using the `set`/ `update` method or using a computed Signal. | Updated by emitting new values. |
| Change detection | Improve change detection performance. Angular can efficiently track changes and re-render when needed. | Using Observables might trigger inefficient change detection. |
| Providing notification | Notify consumers when the hold data changes, facilitating value recalculation or template re-rendering. | Notify consumers when an event occurs or data is emitted, facilitating value recalculation or template re-rendering. |
| Reacting to notification | React to notifications using effects. | React to notifications using callbacks. |

Figure 8.2: Signals versus Observables

But when should you use each one? Well, RxJS shines in scenarios demanding complex reactive data flows. These include the following:

- Managing multiple streams, often arising from asynchronous operations such as HTTP requests
- Handling complex data manipulation such as combination, merging, transforming, and filtering
- Reacting to each emission

On the other hand, Signals are good for the following aspects:

- Simple reactive data management within components, side effects, and calculations
- Data binding scenarios where you want to track changes and trigger targeted UI updates
- Situations where a simpler syntax and potentially improved change detection performance are desired

Signals and RxJS are not mutually exclusive; they can be complementary tools in your Angular development. Angular has several RxJS interop features that make Signals and Observables play nicely in the same app, meaning you can get the benefits of both for a more powerful way to manage your data. These RxJS interop features can be found under the `@angular/core/rxjs-interop` package and include the `toSignal()` and `toObservable()` functions. We'll look at both of these now.

## Understanding the behavior of toSignal()

The `toSignal()` function lets you create a Signal from an Observable. It provides synchronous access to the values that are emitted from the Observable, always containing the most recent emitted values by the Observable. But the coolest part is that `toSignal()` automatically subscribes to the defined Observable and unsubscribes when the component or service that calls `toSignal()` is destroyed.

So, we don't have to manage subscriptions. Doesn't this concept remind you of the async pipe? Indeed; both Signals and the async pipe offer ways to display reactive data in Angular templates. However, Signals provide greater flexibility. Unlike the async pipe, which is primarily used with Observables in templates, Signals can be used anywhere in your application for efficient data management.

But wait – earlier, we learned that a Signal should always have a value while Observables may not emit a value right away. And that's true. That's why `toSignal` has the option to provide an initial value, which will represent the Signal's value until the Observable emits. Here's a simple example:

```
import { toSignal } from '@angular/core/rxjs-interop';
value$ = of([{ name: 'EURO', id: 1 }]);
valueAsSignal = toSignal(this.value$, { initialValue: [] });
constructor() {
effect = effect(() => console.log(this.valueAsSignal()));
}
//console output
{ name: 'EURO', id: 1 }
```

In this example, we created a `value$` Observable using the `of()` creation function, which emits the `[{ name: 'EURO', id: 1 }]` array. Then, we created a Signal named `valueAsSignal` using the `toSignal` function. We pass two arguments to the `toSignal` function:

- `this.value$`: The Observable you want to convert into a Signal.
- `{ initialValue: [] }`: An optional object that allows you to customize the Signal's behavior. Here, we're setting the `initialValue` property to an empty array (`[]`). This ensures that the Signal has a defined value even before the Observable emits its first item.

Finally, we registered an effect to log the Signal's value in the console.

Note that if you don't mention an initial value in the `toSignal` function, the Signal will have `undefined` as the initial value. Be aware that using `undefined` as an initial value can always lead to many errors or inconsistencies, so it would be better to manage this when it's created and provide an initial value.

With that clear, why is this different from using an Observable? Let's focus on another example:

```
import { toSignal } from '@angular/core/rxjs-interop';
values$ = of(10, 20, 30);
this.values$.subscribe(value=> console.log(value));
//console output
10, 20, 30
```

Here, we're creating an Observable that uses the `of` creation function – we subscribe to it and log the values in the console. This Observable emits `10`, `20`, and `30`, respectively, and those values will get logged in the console.

Now, let's convert this Observable into a Signal:

```
import { toSignal } from '@angular/core/rxjs-interop';
values$ = of(10, 20, 30);
valuesAsSignal = toSignal(this.values$, { initialValue: 0 });
Constructor() {
  effect = effect(() =>
    console.log(this.valuesAsSignal()));
}
//console output
30
```

Here, we used the same `values$` Observable and converted it into a Signal using `toSignal` while setting an initial value of 0. Then, we defined an effect to log the value of the Signal. The console output is `30`. Yes, only `30`. Why?

The `of()` creation function emits its values immediately upon subscription. So, when `toSignal` subscribes, all the values are immediately emitted. By the time the effect is scheduled to run, `30` is already in the Signal as the last value emitted and that's what it is logged as to the Signal.

Now, let's delay the `values$` emission by 5 seconds using the `delay` operator:

```
import { toSignal } from '@angular/core/rxjs-interop';
  values$ = of(10, 20, 30).pipe(delay(5));
valuesAsSignal = toSignal(this.values$, { initialValue: 0 });
  effect = effect(() =>
    console.log(this.valuesAsSignal()));
//console output
10, 20, 30
```

When you re-execute the code, you will see 10, 20, and 30 in the console. The effect now has the opportunity to run after each emission because we set up a delay of 5 seconds.

The point here is that when we create Signals, the Signal will not necessarily get notified of all emitted items; it depends on how the Observable is created and on its set of operators.

> **Note**
>
> Signals created through the toSignal() function are read-only – this makes sense as the Signal here is just a consumer of the values emitted by the Observable. Also, keep in mind that toSignal() creates a subscription – you should avoid calling it repeatedly for the same Observable, and instead reuse the Signal it returns.

That is all you need to know about toSignal(). Now, let's explore the toObservable() function.

## Understanding the behavior of toObservable()

If you want to react to a Signal change and perform an async operation such as issuing an HTTP request, the toObservable() function is your friend!

The toObservable() function allows you to convert a Signal into an Observable. Whenever the Signal's value changes, the Observable automatically emits a notification with the new value. This allows you to easily trigger your async operation based on the updated Signal data. Under the hood, toObservable() uses effects to track the Signal's value and emit the latest value to the Observable, as discussed earlier in this chapter.

The toObservable() function might remind you of the asObservable function that's available for subjects, something we explored in *Chapter 7, Sharing Data Between Angular Components*, but these functions don't have the same behavior.

Let's look at an example of the asObservable function first:

```
value = new BehaviorSubject(10);
constructor() {
  this.value.asObservable().pipe(tap(x=>console.log(
    `The value is : ${x}`))).subscribe();
  this.value.next(20);
  this.value.next(30);
}
//console output
The value is : 10
The value is : 20
The value is : 30
```

When using `Subject` or `BehaviorSubject`, notifications are asynchronous. Here, we defined a `BehaviourSubject` subject called `value` with an initial value of `10`. Then, we extracted the `readonly` Observable part of the subject using the `asObservable()` function. Each emitted value is piped through a `tap` operator to log it in the console. Then, we subscribe to start receiving notifications. Finally, in `constructor`, we emit new values (`20` and `30`) using the `next` method.

However, `toObservable` operates differently. It uses an effect where Signal change notifications are scheduled rather than immediately processed, as Observable notifications are. Let's modify the same example by using Signals and `toObservable` instead:

```
value = signal(10);
  constructor() {
    toObservable(this.value).pipe(tap(x=>console.log(
      `The value is : ${x}`))).subscribe();
    this.value.set(20);
    this.value.set(30);
}
//console output
The value is : 30
```

Here, we defined a Signal instead of the `BehaviorSubject` subject named `value` with an initial value of `10`. Then, we called `toObservable(this.value)` to emit a notification when the Signal's value changes. In the pipeline, we once again logged the emitted values and subscribed to the Observable to start receiving notifications. Finally, the value of the Signal was updated using the `set` method.

However, look at the console output – that is, `The value is : 30`. This may not be what you expected, right? This is because the effect behind `toObservable` runs only after the Signal has settled values. The current value of the Signal at that time is the last emitted value, which is `30`.

Keep this behavior in mind when you decide to use a subject or a Signal – a subject will emit all the values from its source, while `toObservable` only emits the current value from the Signal.

> **Note**
>
> Please note that the `toObservable` and `toSignal` functions require an injection context to function properly.

As we delved into the powerful capabilities of both `toSignal` and `toObservable`, you might have noticed the potential for synergy between RxJS and Signals. In the next section, we will learn how we can use both RxJS and Signals in our recipe app and get the best of both worlds.

## Integrating Signals into our recipe app

In this section, we will level up the recipe app's reactive patterns by integrating Signals. We'll kick things off by revisiting the data fetching use case we implemented in *Chapter 3*, and then see how we can adjust the implementation by using Signals in the `RecipesListComponent` to maximize efficiency.

### Fetching data as streams using Signals

Let's briefly review the code snippets we covered for implementing data fetching in `RecipesService` and `RecipesListComponent`.

In `recipes.service.ts`, we have the following code:

```
export class RecipesService {
  recipes$ =
    this.http.get<Recipe[]>(`${BASE_PATH}/recipes`);
  constructor(private http: HttpClient) { }
}
```

In `recipes-list.component.ts`, we have this code:

```
export class RecipesListComponent {
  recipes$ = this.service.recipes$;
  constructor(private service: RecipesService) {
}
```

Finally, in `recipes-list.component.html`, we have this code:

```
@if (recipes$ | async; as recipes) {
// extra code here
}
```

Here, `recipes$` is created in `RecipesService` and represents the Observable that holds the list of recipes. Then, `recipes$` is defined in `RecipesListComponent` and subscribed to in the template using the async pipe. This code snippet was explained in detail in *Chapter 3*.

Now, instead of exposing `recipes$` as an Observable in `RecipesListComponent`, we can consider using a Signal to bind it in the template. To achieve this, we will convert the `recipes$` Observable into a Signal named `recipes` using the `toSignal()` function.

First, to centralize data management in a single place, we will create the `recipes` Signal inside `RecipesService`:

```
import { toSignal } from '@angular/core/rxjs-interop';
export class RecipesService {
  recipes$ =
```

```
    this.http.get<Recipe[]>(`${BASE_PATH}/recipes`);
  recipes=toSignal(this.recipes$, {initialValue: [] as
    Recipe[]});
  constructor(private http: HttpClient) { }
}
```

Here, the `recipes` Signal is created using the `toSignal` function, which takes two arguments:

- `This.recipes$`: The Observable to be converted into a Signal.
- `{initialValue: [] as Recipe[]}`: This is an optional configuration object that specifies an initial value of an empty array, `[]`. This ensures the Signal always has a value, even before the Observable emits any data. We used the TypeScript `as` assertion to define the type of `Recipe[]`.

> **Note**
>
> We can optimize the `RecipesService` code by deleting the `recipes$` property and including its result in the `recipes` property, as follows:
>
> ```
> recipes = toSignal(this.http.get<Recipe[]>(`${BASE_PATH}/
> recipes`), { initialValue: [] as Recipe[] });
> ```

Next, in `RecipesListComponent`, we will define the Signal we created in `RecipesService`:

```
export class RecipesListComponent {
recipes = this.service.recipes;
constructor(private service: RecipesService) {}}
```

Finally, as `toSignal` automatically subscribes to the `recipes$` Observable, we will change `recipes$ |async` to `recipes()` in the `RecipesListComponent` template so that it reads the Signal's value:

```
@if (recipes(); as recipes) {
// extra code here
}
```

No other changes are required. If you go to the app, the list is displayed, and our app still works. By doing this, we kept the Observable-based logic in `RecipesService` for managing async operations using the HTTP client and then created a Signal from that Observable for use in the template. By doing this, we can improve change detection in the template.

Now, how do we handle errors in our Signals? If they are just simple containers of values, how can they generate an error?

In `RecipesService`, we handled the error using the `catchError` operator (discussed in *Chapter 4, Handling Errors Reactively*) and provided a replacement Observable:

```
recipes$.pipe(catchError(() => of([])));
```

This code works fine when using `toSignal`. It is an option to handle errors at the Observable level so that when the Observable that's used in `toSignal` throws an error, this is later caught using the `catchError` operator, and a replacement Observable is provided.

However, if an Observable called in `toSignal` rethrows an error and doesn't handle it (the catch and rethrow strategy detailed in *Chapter 4*), then this error will be thrown each time the Signal is read. Consequently, if the Signal is read multiple times, the error will be thrown repeatedly.

Therefore, if you intend to rethrow the error and perform actions such as displaying a popup message in the UI, then it is highly recommended to catch the error at the Observable level and return an error object as a value. Here's an example:

```
observable$.pipe(
    catchError((error: HttpErrorResponse) =>of({ status: 'error',
description: error })));
```

Here, we have an Observable that catches errors of the `HttpErrorResponse` type and returns an object containing the status (indicating whether it's an error or success) and the error description. At this point, you can register an effect to handle this error at the component level.

Another option is to reject errors completely using the `rejectErrors` parameter of `toSignal`:

```
    recipes = toSignal(this.http.get<Recipe[]>(`${BASE_PATH}/recipes`),
{ initialValue: [] as Recipe[], rejectErrors:true });
```

When enabled, errors are thrown back into the Observable and will become uncaught exceptions. You can imagine `toSignal` saying, "I don't want your errors; take them back." You can then register a global error handler to handle uncaught exceptions and perform your actions:

```
export class GlobalErrorHandler implements ErrorHandler {
    handleError(error: any): void {
      alert(error.message);
    }
}
```

> **Note**
>
> If an Observable that's used in `toSignal` completes, the Signal continues to return the most recently emitted value before completion.

Now that we've used `toSignal` to improve our implementation and understood its behavior in handling errors, along with the various recommended options available, let's circle back to the concept of filtering streams, a topic we explored in *Chapter 5, Combining Streams*. We will use computed Signals to meet the filtering requirement using RxJS and Signals.

## Combining streams using Signals

In the recipe app, we implemented filtering using `BehaviorSubjects`, which effectively notifies components when the filter changes to refine the results. However, Signals also offer a mechanism to react to value changes. They can trigger actions within effects or computed Signals.

This functionality overlaps somewhat with `BehaviorSubjects`, which begs the question, can we replace `BehaviorSubjects` with Signals to filter streams? Let's refresh our memory on the code provided in *Chapter 5*.

In `recipes.service.ts`, we have the following code:

```
export class RecipesService {
  recipes$ =
    this.http.get<Recipe[]>(`${BASE_PATH}/recipes`);
  private filterRecipeSubject = new
    BehaviorSubject<Recipe>({ title: '' });
  filterRecipesAction$ =
    this.filterRecipeSubject.asObservable();
  constructor(private http: HttpClient) { }
  updateFilter(criteria: Recipe) {
    this.filterRecipeSubject.next(criteria);
  }
}
```

In `recipes-list.component.ts`, we have this code:

```
export class RecipesListComponent {
  recipes$ = this.service.recipes$;
  filterRecipesAction$ = this.service.filterRecipesAction$;
  filteredRecipes$ = combineLatest([this.recipes$,
    this.filterRecipesAction$]).pipe(
    map(([recipes, filter]: [Recipe[], Recipe]) => {
    const filterTitle = filter?.title?.toLowerCase() ?? '';
    return recipes.filter(recipe =>
    recipe.title?.toLowerCase() .includes(filterTitle))
  })
  );
  constructor(private service: RecipesService) {
}}
```

Here, `filterRecipesAction$` is the Observable that holds the latest filter's value. It's defined in `RecipesService` and used in `RecipesListComponent` to refine the search. The filter's value is updated through the `updateFilter` method by `RecipesFilterComponent`. `filteredRecipes$` represents the result of filtering; we subscribed to it in the `RecipesListComponent` template using the async pipe. This code snippet is explained in detail in *Chapter 5*.

Now, using Signals, we can replace `BehaviorSubject` and the Observable we created in `RecipesService` with a single Signal named `filterRecipe` and initialize it with an empty value:

```
export class RecipesService {
  recipes =
    toSignal(this.http.get<Recipe[]>(
    `${BASE_PATH}/recipes`), { initialValue: [] as Recipe[]
    });
  filterRecipe = signal({ title: '' } as Recipe);
  constructor(private http: HttpClient) { }
  updateFilter(criteria: Recipe) {
    this.filterRecipe.set(criteria);
}}
```

Here, we created the `filterRecipe` Signal and initialized it with an empty criteria. In the `updateFilter` method, which is used to notify the behavior subject of the change, we will simply update the value of the Signal using the `set` method.

Then, in `RecipesListComponent`, instead of combining streams using `combineLatest`, we will create a computed Signal that will return an array of recipes based on the Signal's filter and the Signal's recipes list. Then, we will refine the recipes list given the filter value using the same filtering function:

```
export class RecipesListComponent {
  recipes = this.service.recipes;
  recipesFilter = this.service.filterRecipe;

  filteredRecipes = computed(() => {
    const filterTitle =
      this.recipesFilter()?.title?.toLowerCase() ?? '';
    return this.recipes().filter(recipe =>
      recipe.title?.toLowerCase()
      .includes(filterTitle));
  })
  constructor(private service: RecipesService) {
  }
}
```

Finally, in the `RecipesListComponent` template, we will be removing the async pipe and replacing it with the call to the `filteredRecipes` Signal, as follows:

```
@if (filteredRecipes(); as recipes) {
    // Extra code here// Extra code here
}
```

This way, we have much cleaner code and an enhanced change detection mechanism.

We used `BehaviorSubjects` in *Chapter 7*, to share the last selected recipe from `RecipesList` `Component` throughout the entire recipe app. Then, we consumed the last shared selected recipe and displayed its details in `RecipeDetailsComponent`. Let's use Signals for the same purpose in this implementation.

## Sharing data using Signals

Before diving into using Signals, let's review the steps that were covered in *Chapter 7*.

In the `shared-data.service.ts` file, we have the following code:

```
export class SharedDataService {
private selectedRecipeSubject = new
BehaviorSubject<Recipe>({});selectedRecipeAction$ = this.
selectedRecipeSubject.asObservable();updateSelectedRecipe(recipe:
Recipe) { this.selectedRecipeSubject.next(recipe);
    }
}
```

In the `recipe-details.component.ts` file, we have this code:

```
export class RecipeDetailsComponent {
constructor(private sharedService: SharedDataService) { }
selectedRecipe$ = this.sharedService.selectedRecipeAction$;
}
```

And in the `recipe-details.component.html` file, we have this code:

```
@if (selectedRecipe$ | async; as recipe) {
}
```

`selectedRecipeAction$` is the Observable that holds the latest selected recipe. It's defined in `SharedDataService` and used in `RecipeDetailsComponent` to display the details. The last selected recipe is updated through the `updateSelectedRecipe` method by `RecipeListComponent`. Then, we subscribed to `selectedRecipe$` in the template using the async pipe. This code snippet was explained in detail in *Chapter 7*.

Now, we'll switch from `BehaviorSubject` to Signals in `SharedDataService`. We'll initialize the created Signal, `selectedRecipe`, with an empty object and change the `updateSelectedRecipe` method so that it updates the value stored in the `selectedRecipe` Signal using the `set` method:

```
export class SharedDataService {
  selectedRecipe = signal({} as Recipe);
  updateSelectedRecipe(recipe: Recipe) {
    this.selectedRecipe.set(recipe);
  }
}
```

So far, so good – we have a Signal that will always hold the last selected recipe.

Next, let's consume this Signal's value in `RecipeDetailsComponent`. We will start by defining the Signal created in `SharedDataService`, as follows:

```
export class RecipeDetailsComponent {
  constructor(private sharedService: SharedDataService) { }
  selectedRecipe = this.sharedService.selectedRecipe;
}
```

Then, in the template, replace `selectedRecipe$ | async` with `selectedRecipe()` to read the value of the Signal.

And we're done. When running this code, you'll notice that the functionality remains intact. Each time a recipe is selected from the list, `RecipeDetailsComponent` will display its details. Now. let's use Signals and `toObservable` to fetch a specific recipe from the server.

## Transforming streams using Signals

Considering the previous example, the recipes array that's displayed in `RecipesListComponent` already contains all the recipe objects, along with their details, so we simply used the client-side recipe object when clicking on a recipe from the list.

Now, imagine that we need to dynamically fetch a recipe's details based on its ID, from a backend service with the `/api/recipes/:recipeID` endpoint (this service is implemented in our `recipes-book-api` backend server; the code is available in this book's GitHub repository). Here's how we can adapt our previous implementation to handle this use case.

We can keep using Signals to track the currently selected recipe's ID. So, in `SharedDataService`, we'll adjust our implementation as follows:

```
export class SharedDataService {
  selectedRecipeId = signal<number | undefined>(undefined);
  updateSelectedRecipe(recipeId: number | undefined) {
    this.selectedRecipeId.set(recipeId);
```

```
    }
}
```

Here, we defined a Signal named `selectedRecipeId` that's been initialized to `undefined` as we don't have an initial selection.

The `updateSelectedRecipe` method now takes `recipeId` (either a number or undefined) as input and updates the `selectedRecipeId` Signal using the `set` method.

Now, in `RecipeListComponent`, we will update the `editRecipe` method so that it only sends the recipe's identifier instead of the whole recipe object:

```
editRecipe(recipe: Recipe) {
    this.sharedService.updateSelectedRecipe(recipe.id);
    this.router.navigate(['/recipes/details']);
}
```

Now, we need to issue an asynchronous HTTP request to fetch the recipe's details whenever a recipe is selected from the list. Observables are ideal for this process! As we learned in *Chapter 6, Transforming Streams*, we need a higher-order mapping operator that does the following:

- Transforms each emitted recipe's identifier into a new Observable that issues an HTTP request
- Cancels the previous HTTP request when a new recipe's identifier arrives and switches to the newly created HTTP request for the latest ID

You may have guessed already, but `switchMap` is the ideal operator to use here.

But wait! We need two key streams involved in this situation:

- *A HTTP request stream*: This stream, which is created using `this.http.get<Recipe>(`${BASE_PATH}/recipes/${id}`)`, represents the actual HTTP request to retrieve the recipe data based on the provided ID.
- *A selected recipe ID stream*: Currently, the selected recipe ID is stored in the `selectedRecipeId` Signal. Here, we can use the `toObservable` function to convert the `selectedRecipeId` Signal into an Observable stream that will emit a notification whenever the selected recipe ID changes in the Signal. It will look like this: `toObservable(this.selectedRecipeId)`.

Now, using the `switchMap` operator, we will define the `recipe$` stream in `SharedDataService`, as follows:

```
recipe$ =
  toObservable(this.selectedRecipeId).pipe(filter(
  Boolean), switchMap(id =>
  this.http.get<Recipe>(`${BASE_PATH}/recipes/${id}`)
));
```

The resulting Observable, `recipe$`, represents a specific recipe object stream. It emits a new recipe whenever the selected recipe ID changes and a successful HTTP request is made.

Finally, within `RecipeDetailsComponent`, we can convert the `recipe$` Observable back into a signal using the `toSignal` function:

```
selectedRecipe = toSignal(this.sharedService.recipe$);
```

This allows us to bind the recipe's data in the component's template using Signals.

Awesome, right? This transformation pattern of using Signals is applicable for every similar use case where you need to combine or transform multiple data streams in your Angular applications!

By leveraging both Angular Signals and RxJS, you can achieve a well-balanced approach to reactive data management in your Angular applications. This harmonious blend allows you to build highly dynamic and responsive user interfaces. Now, let's delve into some interesting new features regarding reactive data binding with Signals.

## Exploring reactive data binding with Signals

Angular's data binding capabilities have been steadily improving to support reactivity. Starting from version 17.1, Angular introduced some powerful features to leverage reactivity using Signals in component interaction and data binding, such as input Signals, model inputs (starting from 17.2), and support for content and view queries. To align with input Signals, version 17.3 provides a new output API.

We will explore these new features in this section.

### Signal inputs

The Angular `@Input` decorator is used to define an input property in a component, allowing data to be passed into the component from its parent component or template. It essentially creates a one-way data flow from the parent to the child component.

Angular 17.1 introduces Signal inputs that allow input data to be passed as Signals. This adds a powerful twist to data binding between parent and child components in Angular, transforming traditional Angular inputs into a reactive data source. Here's an example:

```TypeScript
@Component({
  selector: 'app-shipping',
})
export class ShippingComponent {
  addressLine2 = input<string>();
  identifier = input(0);
```

```
    addressLine1 = input.required<string>();
}
```

In this example, we defined three Signal inputs:

- `addressLine2`: An optional input that can hold a string value or be undefined.

- `identifier`: An optional input that holds a number and has a default value of 0.

- `AddressLine1`: A required input that holds a string value. It is declared using the `input.` `required` function, and by default, the inputs are optional (that's why Signal inputs are type-safe). If not provided, a compilation error will be thrown, like so: **NG8008: Required input '** **addressLine1' from component ShippingComponent must be specified**.

> **Note**
>
> Required inputs cannot have a default value. Therefore, you can't read their values before they've been bound, and Angular throws an exception. Consequently, you can't access their values in the constructor. However, you can safely access the values within ngOnInit, ngOnChanges, computed, or effects as they are only triggered when the component has been initialized.

> **Note**
>
> When referenced in templates, Signal inputs will automatically mark OnPush components as dirty.

Now that we've got a handle on creating Signal inputs and understanding their syntax, you might be curious about how to use them. Signal inputs are read-only. You can access the value by calling the getter function in the template, as follows:

```
{{addressLine1()}}
{{addressLine2()}}
```

You can also bind to an input Signal like so:

```
< app-shipping addressLine1 ="2300 Vision Lane">
< app-shipping [addressLine1]="addressProperty">
< app-shipping [label]="addressAsSignalProperty()">
```

In this example, we bound the Signal input property, `addressLine1`, to different values: a string named `2300 Vision Lane`, a component property named `addressProperty`, and a Signal's value named `addressAsSignalProperty()`.

Binding to Signals opens the door to a whole new level of dynamic data flow; any changes made to the input value in the parent component will be automatically reflected in the child component. This is where the real magic happens. In the following example, we're using the Signal input property's name to bind the values, but you can provide an alias to the input name using the following syntax:

```
identifier = input(0,{alias: 'id'});
```

This allows you to reference the input by using `<app-shipping [id]=50>` as the alias in the template while still using `this.identifier` as the property name inside your component.

In addition to using signal inputs for value binding in templates, they can also be used within `effects` and `computed` functions. Are you wondering how to do that? Let's look at some examples.

Here, we're appending the values of `addressLine1` and `addressLine2` in the computed function to build `fullAddress`:

```
fullAddress = computed(() => `${this.addressLine1()}
    ${this.addressLine2()}` );
```

It's possible to track the Signal input changes using the `effect` function, like so:

```
constructor() {
    effect(() => {
        console.log(this.identifier());
    });
}
```

In this example, the `console.log` function is invoked every time the identifier input changes. This is a new way to track value changes.

So, life cycle hooks such as `ngOnInit` and `ngOnChanges` can now be replaced with `computed` and `effect`, making value monitoring easier. Instead of implementing extra code inside `ngOnInit` or `ngOnChanges`, we can simply register `effect` to monitor values and use `computed` to perform automatic calculations.

With that, we've covered the essentials of Signal inputs, which enable one-way data binding. Next, we'll explore how bidirectional data binding can be achieved using Signals.

## Model inputs

Model inputs are similar to the previously explained Signal inputs, allowing you to bind a value into a property. However, model inputs allow the component to write values into the property, unlike other inputs, which are read-only. This enables two-way reactive data binding, allowing the child component to not only receive data changes from the parent but also notify the parent of any changes it makes to the data.

Let's look at an example. Here's some TypeScript code:

```
identifier = model(0,{alias: 'id'});;
constructor() {
  setInterval(() => {
    this. identifier.set("000524");
  }, 4000)
}
```

And here's some code in the HTML template:

```
<app-shipping [(id)]=counter></ app-shipping>
{{counter()}}
```

In the TypeScript code, we converted the Signal input named `identifier` into a model input that initially contains a value of 0. Then, in the constructor, we set up a timer that will update the value of the `identifier` input after 4 seconds.

Then, in the HTML template, we simply used the two-way data binding syntax to bind to a property called `counter` that we defined in the parent component and then display the `counter` value.

When running this code, you will see that the model's input value will get updated after 4 seconds to `000524` and the counter property will have `000524` as its value as well. The parent component is automatically notified.

Another thing to note when defining a model input is that, under the hood, Angular generates an output for that model. The output's name is just the model input's name suffixed with `Change`:

```
<app-shipping [(id)]=counter (idChange)="updateMessage()" ></ app-
shipping>
```

Here, we called the `idChange` output and triggered the `updateMessage`that method, which will display an alert when the model value changes. The `idChange` event will be emitted whenever you write a new value into the model input.

## Signal queries

Signal queries present a reactive alternative to traditional queries declared with the `@ContentChild`, `@ContentChildren`, `@ViewChild`, or `@ViewChildren` decorators. Signal queries expose query results as Signals, which means that query results can be composed with other Signals (using `computed` or `effect`) and drive change detection.

For more details, you can check out the official documentation: `https://angular.dev/guide/signals/queries`.

## Summary

This chapter took a deep dive into Angular Signals. We started by figuring out why Signals exist and how they help manage data reactively.

Then, we explored the Signals API, from creating and reading the current value to using computed Signals and effects when the value changes.

Next, we compared Signals to RxJS Observables. We saw what each is good at and when you'd use one over the other. Angular even provides special interop functions that let Signals and Observables work together nicely, including `toObservable()` and `toSignals()`, both of which we discussed.

Finally, to put everything into practice, we used Signals in our recipe app to see how they work with RxJS in real-world scenarios. This hands-on experience helped us solidify what we learned about using Signals and RxJS together. We also walked through the newest improvements regarding reactive data binding and component interaction using Angular Signals.

By incorporating Angular Signals into your Angular applications, you can streamline data management, enhance code readability, and leverage the power of reactive programming. Remember, signals and RxJS work together to empower you to build dynamic and responsive user interfaces.

In the next chapter, we'll move on to the essentials of multicasting, which will be helpful in the following chapters.

# Part 4:
# Multicasting Adventures

In this part, we will understand the essentials of multicasting in RxJS, as well as the recommended reactive patterns in many real-world use cases, such as caching data, multiple asynchronous operations, and real-time features.

You will also delve into the best practices when using multicast operators, Subjects, and Behavior Subjects and learn the pitfalls to avoid specifically in the context of multicasting.

This part includes the following chapters:

- *Chapter 9, Demystifying Multicasting*
- *Chapter 10, Boosting Performance with Reactive Caching*
- *Chapter 11, Performing Bulk Operations*
- *Chapter 12, Processing Real-Time Updates*

# 9
# Demystifying Multicasting

**Multicasting** refers to sharing the same Observable execution among multiple subscribers. This concept can be challenging to grasp initially, especially for those unfamiliar with reactive programming paradigms. However, it's very useful and solves many problems in web applications.

In this chapter, I will demystify this concept, explaining when and where to use it, how RxJS subjects are involved, and its advantages.

So, in this chapter, we're going to cover the following main topics:

- Explaining multicasting versus unicasting
- Exploring RxJS subjects
- Highlighting the advantages of multicasting

## Technical requirements

This chapter assumes that you have a basic understanding of RxJS.

All the source code in this chapter is used for demonstration purposes, so you don't need access to this book's GitHub repository.

## Explaining multicasting versus unicasting

Before we get into explaining multicasting versus unicasting, let's start by explaining another key concept, known as a producer, which we will be using a lot in this chapter.

A **producer** is the source that produces the Observable values – for example, DOM events, WebSockets, and HTTP requests are considered producers. It is any data source that's used to get values.

Observables fall into two types:

- Cold, or unicast, Observables
- Hot, or multicast, Observables

Let's understand the difference between them.

## Unicasting and cold Observables

A cold Observable in RxJS is like a personal storytelling session. Imagine you're sharing a story with a friend. You narrate the story right there with them, and it's unique to your interaction. Each time you share the story with a different friend, it's like starting a new session with a fresh narrative.

In RxJS terms, this means that the Observable itself generates the data it emits. Each time someone subscribes to the Observable, they get a private storytelling session. The story (or data) isn't shared between different listeners – it's a one-on-one experience. This is why we call cold Observables "unicast" – each emitted value is observed by only one subscriber:

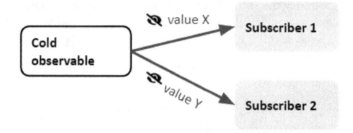

Figure 9.1 – Unicast cold Observable

So, by default, Observables in RxJS are cold – they create and deliver data to each subscriber individually, just like your personalized storytelling sessions.

Here's an example of a cold Observable:

```
import { Observable} from 'rxjs';

const coldObservable$ = new Observable(observer => {
  observer.next(Math.random());
  observer.next(Math.random());
  observer.complete();
});

/** First subscriber */
coldObservable$.subscribe(data => {
  console.log(`The first Observer : ${data}`);
});

/** Second subscriber */
coldObservable$.subscribe(data => {
```

```
  console.log(`The second Observer : ${data}`);
});

//console output
The first Observer: 0.043246216289945405
The first Observer: 0.7836505017199451
The second Observer: 0.18013755537578624
The second Observer: 0.23196314531696882
```

Let's break down what's happening in this code.

Here, `Math.random()` is our producer – it is called inside the Observable. So, data is produced by the Observable itself.

The first ubscriber will get two random values after the subscription, and the second subscriber will get two different values after the subscription. Every subscriber starts a new execution, leading to a new invocation of `Math.random()`, which results in distinct values.

Each subscriber gets its own unique set of items. It begins to emit items only after the observer subscribes to it. Since there are two different executions, every Observable will receive a different value. This means that data is unicast and not shared among the subscribers.

Briefly looking at a real-world example, when a user logs into an application, their personal profile or dashboard information is fetched and displayed. This data is unique to each user and should not be shared across multiple users. Using a cold Observable ensures that each user receives personalized data upon logging in, maintaining privacy and security. So, to summarize, for cold Observables, the following applies:

- The Observable itself generates the data it emits
- It starts to emit data only after the observer subscribes to it
- Each observer (or subscriber) gets its own unique set of items

Now, let's look at hot Observables.

## Multicasting and hot Observables

Multicasting in RxJS is like hosting a live radio show. Imagine you're broadcasting a show from a studio, and listeners can tune in at any time to hear the same content. Once you start broadcasting, anyone who tunes in can hear the same music, interviews, or discussions.

In RxJS terms, a hot or multicast Observable is an Observable whose emitted values are shared among subscribers. There's a single source of data, just like the radio station broadcasting content. When you subscribe to a multicasting Observable, you're joining the "broadcast," and you'll receive the same data as anyone else who's tuned in:

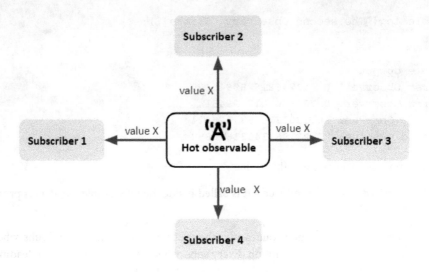

Figure 9.2 – Multicast hot Observable

Unlike cold Observables, where each subscriber gets a private session, multicasting allows multiple subscribers to listen to the same stream of data simultaneously.

Here's an example of a hot Observable:

```
import { Observable, fromEvent } from 'rxjs';
// Hot Observable
const hotObservable$ = fromEvent(document, 'click');

hotObservable$.subscribe(({ x, y }: MouseEvent) => {
  console.log(`The first subscriber: [${x}, ${y}]`);
});

hotObservable$.subscribe(({ x, y }: MouseEvent)=> {
  console.log(`The second subscriber: [${x}, ${y}]`);
});

//console output
The first subscriber: [108, 104]
The second subscriber: [108, 104]
```

Let's break down what's happening in this code.

We created an Observable using the fromEvent function of RxJS. This Observable will emit clicks happening on the DOM document when subscribing to.

> **Note**
>
> For more details about `fromEvent`, please refer to `https://rxjs.dev/api/index/function/fromEvent`.

In this case, the data is emitted outside the Observable and, as you may have guessed, both subscribers will get the same data. This means that the subscribers share the same instance of the DOM click event. So, the hot Observable shares data between multiple subscribers. We call this behavior multicasting. In other words, the Observable multicasts to all subscribers.

Looking at another real-world scenario, consider a chat application where you might have a global chat service that exposes a hot Observable representing the stream of incoming messages from all users in the chat room. Multiple components, such as message feeds and notifications, can subscribe to this hot Observable to display new messages in real-time without needing to create separate Observables for each component.

So, to summarize, for hot Observables, the following applies:

- Data is produced outside the Observable

- It may begin emitting items as soon as it is created

- The emitted items are shared between the subscribers (multicasting)

## Transforming cold Observables into hot Observables

If we want to transform the cold Observable into a hot one, we have to move the producer outside the Observable – this way, our subscribers will receive the same data.

Let's revisit our example with the cold Observable. Instead of generating values within the Observable, we'll pre-calculate the value by using `Math.random()` outside the Observable, like this:

```
const value = Math.random();
const coldObservable$ = new Observable(observer => {
  observer.next(value);
  observer.next(value);
  observer.complete();
});

/** first subscriber */
coldObservable$.subscribe(data => {
  console.log(`The first subscriber: ${data}`);
});

/** second subscriber */
coldObservable$.subscribe(data => {
```

```
    console.log(`The second subscriber: ${data}`);
});

//console output
The first subscriber: 0.6642828154026537
The first subscriber: 0.6642828154026537
The second subscriber: 0.6642828154026537
The second subscriber: 0.6642828154026537
```

As you may have noticed, after executing this code, all subscribers receive the same pre-calculated value.

Now, before we wrap up this section, let's just quickly summarize unicasting and multicasting:

- You should use *unicasting* when you want each subscriber to own independent executions and separate data streams for each subscriber.

- On the other hand, you should use *multicasting* when you want to make sure multiple subscribers share the same execution and results, particularly in scenarios involving hot Observables, broadcasting, or caching results.

  Multicasting also helps optimize and improve performance when executing data is expensive. As a quick final example, suppose that the Observable's execution is issuing a network request. If we choose a cold Observable (or unicasting), then a network request will be raised for every subscriber. Instead, multicasting is a better fit for this particular scenario as it will share the execution of the network requests among subscribers and consequently avoid redundant request calls.

Now that we understand multicasting and hot Observables, let's explore the most popular ways to multicast values to observers in RxJS, namely RxJS subjects.

## Exploring RxJS subjects

**Subjects** are special types of Observables. While plain Observables are unicast, subjects are multicast, allowing values to be broadcast to all subscribers.

You can consider subjects as observers and Observables at the same time:

- You can subscribe to subjects to get values emitted by the producer (that's why they act as Observables):

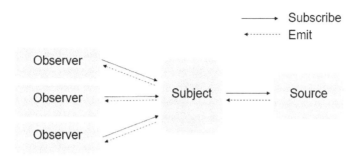

Figure 9.3 – An RxJS subject

- You can send values, errors, and completes by using the `next`, `error`, and `complete` methods that are available in the `Observer` interface (that's why they act as observers):

```
const observer = {
  next: x => console.log('Observer got a next value: '
                         + x),
  error: err => console.error('Observer got an error:
                              '+err),
  complete: () => console.log('Observer got a
                             completion'),
};
```

In short, a subject maintains a list of subscribers and notifies them when a new value is emitted. But to go a bit deeper, there are multiple types of subjects in RxJS. Let's explore the most used ones.

## A plain subject

`plainSubject` is the parent type of all subjects. Let's see a quick example:

```
const plainSubject$ = new Subject();
plainSubject$.next(10);
plainSubject$.next(20);

plainSubject$.subscribe({
    next: (message) => console.log(message),
    error: (error) => console.log(error),
    complete: () => console.log('Stream Completed'),
  });

plainSubject$.subscribe({
    next: (message) => console.log(message),
    error: (error) => console.log(error),
```

```
    complete: () => console.log('Stream Completed'),
  });


plainSubject$.next(30);

//console output
30
30
```

In the preceding code, we created `plainSubject$`, which emitted 10 and 20 as values. Afterward, we created two subscribers that logged the incoming values, the errors, and the completion of the stream. Finally, `plainSubject$` emitted a value of 30.

After executing this code, notice that only 30 is traced twice in the console. This means that the subscribers only received 30. Why have they not received 10 and 20? Because those values were emitted before the subscription to `plainSubject$` and every emission that occurs before the subscription will get lost. This is how a regular subject multicasts values.

And that's how plain subjects behave and emit values.

You can use subjects as a communication hub in your web application to share data between different Angular components, as we explored in *Chapter 7, Sharing Data between Angular Components*.

Moreover, subjects can be used to manage the authentication state in a web application. For instance, you can use a subject to emit a value whenever the user logs in or logs out. This emitted value can then be used to conditionally display certain components or trigger specific behaviors based on the user's authentication status.

If you want to keep a buffer of previous values emitted to subscribers coming late to the game, then `ReplaySubject` can help!

## replaySubject

`replaySubject` is a subject variant, similar to `plainSubject`, but with an in-memory feature: they remember and replay previous messages to new subscribers. Replay subjects have a memory.

Let's explain how it works by exploring the following example:

```
const replaySubject$ = new ReplaySubject();
replaySubject$.next(10);
replaySubject$.next(20);
replaySubject$.next(50);
replaySubject$.subscribe({
  next: (message) => console.log(message),
  error: (error) => console.log(error),
```

```
  complete: () => console.log('Stream Completed'),
});

replaySubject$.subscribe({
  next: (message) => console.log(message),
  error: (error) => console.log(error),
  complete: () => console.log('Stream Completed'),
});
replaySubject$.next(30);

//console output
10
20
50
10
20
50
30
30
```

As you can see, all the values were replayed to the new subscribers. Now, to control the buffer size (the number of values you want the `Replay` subject to store), you can pass it as a parameter when creating `ReplaySubject`, like so:

```
const  replaySubject$ = new ReplaySubject(2);
```

This will only replay the last two values. The console output will be as follows:

```
20
50
20
50
30
30
```

As a real-world use case, let's consider a chat application where a user joins the chat room late. With `ReplaySubject`, they can still see previous messages that were sent before they joined. This is useful for providing a complete chat history to new users.

With that, let's move on to the other variant of `Subject` – `BehaviorSubject`.

## BehaviorSubject

`BehaviorSubject` is just `ReplaySubject` with a buffer size equal to one, so it can only replay only previous item. We used `BehaviorSubject` in *Chapter 5, Combining Streams*.

BehaviorSubject requires an initial value and always retains the last value so that it can emit it to new subscribers. In other words, if you have any subscribers coming late to the game, they will get the previous value that was emitted by the stream. This will always give you value when you subscribe.

Here's an example:

```
const behaviourSubject$ = new BehaviorSubject(1);
behaviourSubject$.next(10);
behaviourSubject$.next(20);
behaviourSubject$.next(50);

behaviourSubject$.subscribe({
  next: (message) => console.log(message),
  error: (error) => console.log(error),
  complete: () => console.log('Stream Completed'),
});

behaviourSubject$.subscribe({
  next: (message) => console.log(message),
  error: (error) => console.log(error),
  complete: () => console.log('Stream Completed'),
});
behaviourSubject$.next(30);

//console output
50
50
30
30
```

Here, behaviourSubject$ is created and has an initial value of 1. Then, behaviourSubject emitted 10, 20, and 50, respectively. Right after we subscribed two times to behaviourSubject$, both subscribers will immediately receive the last value emitted by behaviourSubject$, which is 50 – that's why 50 is traced two times in the console. Finally, behaviourSubject$ emitted 30; consequently, the subscribers will receive 30 and trace it.

If no values were emitted before the subscription, then behaviourSubject$ will emit the initial value, which is 1:

```
const behaviourSubject$ = new BehaviorSubject(1);

behaviourSubject$.subscribe({
  next: (message) => console.log(message),
  error: (error) => console.log(error),
  complete: () => console.log('Stream Completed'),
```

```
});

behaviourSubject$.subscribe({
  next: (message) => console.log(message),
  error: (error) => console.log(error),
  complete: () => console.log('Stream Completed'),
});
behaviourSubject$.next(30);
//console output
1
1
30
30
```

As another example, imagine that you're building a weather app that displays the current temperature. You can use `BehaviorSubject` to represent the temperature data. Whenever the temperature changes, you update `BehaviorSubject` with the new value. Subscribers to `BehaviorSubject` will always receive the latest temperature, even if they start using the app after the temperature has changed multiple times.

To summarize, `PlainSubject`, `BehaviorSubject`, and `ReplaySubject` are the most used subjects in RxJS, which is why we discussed them here. However, there are other types of subjects, such as `WebSocketSubject`, which are used much less (though we will explore that one more in *Chapter 12, Processing Real-Time Updates*). For details about the other types, refer to `https://rxjs.dev/guide/subject`.

---

**Note**

There are also many useful RxJS operators for multicasting (or sharing values/executions) in RxJS 6, namely `multicast`, `publish`, `share`, `shareReplay`, `publishReplay`, `publishLast`, and `refcount`. For more details about these operators, you can check the official docs: `https://rxjs.dev/api/operators`.

In version 7, multicasting operators were consolidated to `share`, `connectable`, and `connect`. The other multicasting APIs are now deprecated and will be deleted in RxJS 8. The only operator that wasn't deprecated is `shareReplay` because it is very popular. It is now a wrapper around the highly configurable `share` operator.

Since we are using RxJS 7 in this book, I think it is useless to go through all the deprecated operators. Instead, we focus on the `share` operator as it satisfies most cases. We will learn the behavior of the `share` operator by considering a real-world use case in the next chapter, *Chapter 10, Boosting Performance with Reactive Caching*.

For details about RxJS 7 multicasting operators, refer to `https://rxjs.dev/deprecations/multicasting`.

Now that you have a good understanding of multicasting and the different ways of implementation provided by RxJS, let's explore the main advantages of multicasting.

## Highlighting the advantages of multicasting

Multicasting in RxJS or having the ability to share the same Observable execution among multiple subscribers has many advantages. Here are some key ones:

- *Optimizing resources*: Multicasting helps optimize resources by avoiding redundant treatments. When dealing with expensive operations such as raising HTTP networks or performing complex computations, multicasting helps you do the work once and share the results among all subscribers.

- *Consistent data and results*: Multicasting ensures that all subscribers receive the same set of values emitted by the Observable. This can be crucial in scenarios where consistency in data is essential, and you want all subscribers to observe the same data sequence.

- *Broadcasting*: Multicasting gives you the possibility to send, once and for all, the same set of values to multiple subscribers at the same time. This is what we call broadcast, and it is beneficial when you have a complex application with multiple components that need to react to the same set of values.

- *Late subscribers*: Multicasting allows late subscribers to receive the same values as subscribers who joined earlier. This is achieved by using `BehaviorSubject` and `ReplaySubject`, plus some multicast RxJS operators such as the `shareReplay` operator, something we will explain in *Chapter 10, Boosting Performance with Reactive Caching*.

When designing applications with RxJS, multicasting is a powerful mechanism that should be put on the table because it enhances performance, efficiency, consistency, and interaction between the different parts of your app. However, it's essential to use multicasting carefully and be aware of potential pitfalls, depending on your specific use case. Here are some examples:

- *Data alteration*: Multicasting inherently shares the same data stream across all subscribers. If a subscriber modifies the data within its subscription logic (for example, using operators such as `map` or `filter`), this modification can unintentionally affect what other subscribers receive. This can lead to unexpected behavior and debugging challenges.

  For example, imagine a multicast Observable emitting a list of products. A component subscribes and filters the list to only show products with discounts. However, this filtering modifies the original data stream. If another component later subscribes, expecting the complete product list, it will only receive the discounted products due to the unintended modification in the first subscription. That's why keeping the subject private and exposing only the read-only part of the data through the `asObservable()` method is a common and effective practice. This ensures that external components or consumers cannot directly modify the internal state of the subject. Instead, they can only observe the emitted values without interfering with the data stream.

- *Memory leaks*: Unlike unicast Observables, which complete after a single subscriber unsubscribes, multicasting continues emitting data so long as at least one subscriber remains. This can lead to memory leaks if you're not careful about managing subscriptions, especially when dealing with infinite or long-lived Observables.

  For example, imagine a multicast Observable that emits real-time stock prices. If components subscribe to this Observable but don't unsubscribe when they are no longer needed, the Observable will continue emitting, potentially causing memory leaks as references to the Observable and its internal state accumulate.

We will explore other multicasting pitfalls and best practices in the next chapters.

## Summary

In this chapter, I walked you through the most important concepts and vocabulary to understand multicasting. We started by explaining the role of a producer, after which we learned the difference between cold and hot Observables, leading us to the definition of multicasting and unicasting. Then, we explored RxJS subjects, the different types of subjects, and the use cases of each before introducing multicasting operators in RxJS.

In the next chapter, we'll practice all of this in a real-world use case. We will learn how to put an efficient mechanism of caching in place in our Recipe app by using multicasting in RxJS and, more specifically, by combining multicasting operators and subjects.

# Boosting Performance with Reactive Caching

Caching data and assets is one of the most efficient ways to improve the user experience of our web applications. It's a good way to speed up the load times of our web applications and keep the number of network requests to a minimum.

We will start this chapter by defining the caching requirement for our application's client side and looking at its motivation. Then, we will learn how to implement this requirement reactively using RxJS operators. After that, we will describe a better way to do this using the latest features of RxJS 7. Finally, we will highlight another use case of caching streams, which is for side effects.

In this chapter, we're going to cover the following main topics:

- Defining the caching requirement
- Exploring the reactive pattern to cache streams
- Highlighting the use of caching for side effects

## Technical requirements

This chapter assumes that you have a basic understanding of RxJS.

The source code of this chapter is available at `https://github.com/PacktPublishing/Reactive-Patterns-with-RxJS-and-Angular-Signals-Second-Edition/tree/main/Chap10`.

# Defining the caching requirement

As you have learned throughout the previous chapters, the `HTTPClient` module is Observable-based, which means that methods such as `get`, `post`, `put`, and `delete` return an Observable. Subscribing multiple times to this Observable will cause the source Observable to be created repeatedly, hence performing a request on each subscription – as we learned in *Chapter 9, Demystifying Multicasting*, this means it is a cold Observable. This behavior will result in an overhead of HTTP requests, which may decrease the performance of your web applications, especially if the server takes some time to respond.

Reducing HTTP requests by caching the result on the client side is one of the most commonly used techniques to optimize web applications. **Client-side caching** involves storing previously requested data so that you don't raise repetitive requests to the server and harm your application's performance.

Let's picture this with a streaming service scenario. Imagine that you're watching your favorite TV show on a streaming service. When you start watching, the streaming service fetches the episodes from the internet and streams them to your device. Now, let's say you want to rewind a bit and watch a scene again. Instead of fetching the episodes from the internet again, the streaming service has already stored the episodes you've watched in a special memory bank. This memory bank allows you to rewind and rewatch scenes without having to re-fetch them from the internet.

But when should we cache data? When data is shared (used by more than one component in your app) and doesn't change frequently, it makes a lot of sense to cache it and share it among multiple components. For example, the user's profile data is subject to caching. We generally retrieve the user's profile information after they log in, and it won't change during the user's session.

Additionally, reference data, such as lists of countries, currencies, or categories, are subjects for caching. Since this doesn't change frequently, you can cache it and share it among multiple components.

In the case of `RecipesApp`, the `/api/recipes` GET request is called every time the `RecipesList` component is rendered to load the list of recipes. In other words, whenever the user clicks on the recipe app's logo or navigates between `HomeComponent` and `RecipeCreationComponent`, a GET request will be issued, even if the list of recipes hasn't changed.

The following screenshot shows the raised requests in the **Network** tab:

Figure 10.1 – The GET HTTP requests and their overhead

As you may have noticed, all those outgoing requests result from the navigation between HomeComponent and the other components.

In this chapter, we will assume that the list of recipes does not change frequently. In this case, it is useless to request the server on every component's load; it would be better to cache the result and read the data from the cache to enhance the performance and the user experience.

But what if there are new recipes? What about updates? Well, there are two techniques we can utilize to handle updates:

- We could update the cache data after each interval of time to retrieve the most recent version of the data – this technique is called **polling**

- We could place a server push notification to get real-time updates instantly

In this chapter, to understand the caching behavior in RxJS through basic examples, we will keep it simple and implement a client-side cache with and without a refresh capability.

> **Note**
> Though we will cover the polling technique in this chapter, we will cover the second technique in *Chapter 11*, *Processing Real-Time Updates*.

So, without further ado, let's look at how we can implement this.

# Exploring the reactive pattern to cache streams

You'll be glad to know that RxJS ships with a very useful operator to implement a stream caching mechanism – this is the shareReplay multicast operator. Let's take a look.

## The shareReplay operator

In RxJS, shareReplay works similarly to the streaming service memory bank, sharing an Observable's execution with multiple subscribers. When you subscribe to an Observable that uses shareReplay, it fetches the data, just like streaming a show. However, shareReplay caches or remembers the emitted values from the Observable. If you subscribe again later, instead of fetching the data again, it replays the cached values from its memory bank.

This can be useful when you have multiple subscribers to an Observable, but you don't want each subscriber to trigger a new data fetch. Instead, you want them to share the same set of data, like multiple viewers sharing the same TV show episodes. This can improve performance and reduce unnecessary data fetching in your application.

So, in a nutshell, the `shareReplay` operator does the following:

- Shares an Observable's execution with multiple subscribers
- Offers the possibility to replay a specified number of emissions to the subscribers

Now, let's see how we can use the `shareReplay` operator for our requirement.

## Using shareReplay in RecipesApp

Our goal is to cache the list of recipes in our app. This is represented by the `recipes$` stream defined in `RecipesService`, as shown here:

```
export class RecipesService {

recipes$ = this.http.get<Recipe[]>(`${BASE_PATH}/recipes`);

}
```

The `recipes$` stream is initially a cold Observable, meaning the stream's data is re-emitted for every subscriber, resulting in an overhead of HTTP requests. This is not what we want. We want to share the last stream's emission with all subscribers – in other words, we want to transform the cold stream into a hot stream using the `shareReplay` operator, like so:

```
export class RecipesService {

recipes$ =
this.http.get<Recipe[]>(`${BASE_PATH}/recipes`).pipe(
shareReplay(1));
}
```

By passing 1 as an argument, `shareReplay` cached the last emission from `recipes$`.

Now, let's explain the complete data-sharing workflow:

- First, HomeComponent is initialized.
- Then, HomeComponent triggers the rendering of the child component – that is, RecipesListComponent.
- RecipesListComponent loads the recipes$ Observable that's available in RecipeService. It will perform the GET HTTP request to retrieve the list of recipes since this is the first time we have asked for the data.
- Then, the cache will be initialized by the data coming back from the server.

- The next time the data is requested, it will be retrieved from the cache thanks to the shareReplay operator. Under the hood, the shareReplay operator creates a ReplaySubject instance that will replay the emissions of the source Observable with all future subscribers. After the first subscription, it will connect the subject to the source Observable and broadcast all its values.

This is the multicasting concept we explained in *Chapter 9, Demystifying Multicasting*. The next time we request the recipes list, our cache will replay the most recent value and send it to the subscriber. No additional HTTP call is involved. So, when the user leaves the page, it unsubscribes and replays the values from the cache.

The following diagram also illustrates the complete workflow:

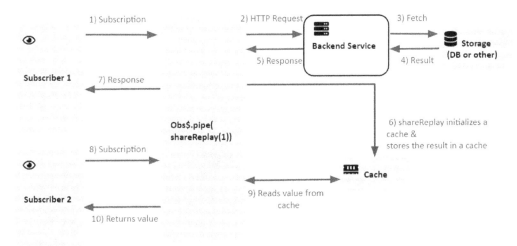

Figure 10.2 – ShareReplay execution

This works perfectly fine when the data doesn't need to be refreshed at all. But as described in the requirement, we need to refresh RecipesList every interval. If the polling technique is used, we can update the cache like so:

```
import { switchMap, shareReplay, timer } from 'rxjs/operators';
const REFRESH_INTERVAL = 50000;
const timer$ = timer(0, REFRESH_INTERVAL);



export class RecipesService {

recipes$ = timer$.pipe(
```

```
    switchMap(_ =>

    this.http.get<Recipe[]>(`${BASE_PATH}/recipes`)),

    shareReplay(1)

  );

}
```

Here, we created a `timer$` Observable that will emit every 50 seconds. This interval is configured in the `REFRESH_INTERVAL` constant, using the `timer` function available in RxJS to create the `timer$` Observable. For more details about the `timer` function, please refer to `https://rxjs.dev/api/index/function/timer#examples`.

Then, for every emission, we use the `switchMap` operator to transform the value into the Observable that's returned by the HTTP client. This will issue an HTTP GET every 50 seconds and, consequently, update the cache.

This is a known RXJS pattern for executing a treatment every *x* seconds.

Now, let's see how we can customize the `shareReplay` operator.

## Customizing the shareReplay operator

With RxJS 6.4.0, a new `shareReplay` signature was provided to customize the operator's behavior. The new signature takes a single `config` parameter of the `ShareReplayConfig` type, as follows:

```
function shareReplay<T>(config: ShareReplayConfig):
MonoTypeOperatorFunction<T>;
```

The `ShareReplayConfig` interface contains the following properties:

```
interface ShareReplayConfig {
  refCount: boolean;
  bufferSize?: number;
  windowTime?: number;
  scheduler?: SchedulerLike;
}
```

Let's discover the purpose of each property:

- `refCount`: If `refCount` is enabled (set to `true`), the `shareReplay` stream will unsubscribe from the source Observable when there are no subscribers. Therefore, the source will no longer emit. This means that if a new subscriber comes along later, then a new stream will be created that subscribes to the source Observable. If `refCount` is disabled (set to `false`), the source will not be unsubscribed, meaning that the inner `ReplaySubject` will still be subscribed to the source and potentially run forever. To avoid memory issues, it is highly recommended to set the `refCount` property to `true`.

- `bufferSize`: This refers to how many values you want to replay. For example, if you just want the one previous value to be replayed for each new subscriber to the shared stream, then you should mention 1 as a `bufferSize` value like so: `shareReplay({bufferSize: 1})`.

- `windowTime`: This refers to the time limit in milliseconds for data stored in the buffer to be emitted to future subscribers.

- `scheduler`: This is used to control the execution and provide a way to manage concurrency (for more details, please refer to the official documentation: `https://rxjs.dev/api/index/interface/SchedulerLike`).

In our case, we need to configure `bufferSize` to 1 to store only the latest value and set `refCount` to `true` to prevent memory leaks.

So, using the `shareReplayConfig` object, the final code of `RecipesService` will look like this:

```
import { switchMap, shareReplay, timer } from 'rxjs/operators';
const REFRESH_INTERVAL = 50000;
const timer$ = timer(0, REFRESH_INTERVAL);
export class RecipesService {

recipes$ = timer$.pipe(
    switchMap(_ =>
    this.http.get<Recipe[]>(`${BASE_PATH}/recipes`)),
    shareReplay({bufferSize: 1, refCount: true })
  );
}
```

When using `shareReplay` on Observables that don't complete on their own, always consider the `refCount` flag.

Now that we know the behavior of `shareReplay`, I want to shed light on an improvement that's become available starting from RxJS 7 that allows you to replace the `shareReplay` operator with the `share` operator.

## Replacing the shareReplay operator with the share operator

The share operator is similar to shareReplay but, by default, it doesn't have a buffer and it doesn't replay that buffer on subscription.

With the share operator, once the subscriber count reaches 0, the source Observable automatically unsubscribes. On the other hand, when the refCount option of shareReplay is set to true, it behaves similarly to the share operator in terms of reference counting, but it also offers the ability to replay emitted values.

Here's a table that compares the two:

| Feature | share | shareReplay |
|---|---|---|
| **Behavior** | Creates a multicast Observable. | Creates a multicast Observable. |
| **Replaying** | Does not replay the previous emission.<br><br>It uses Subjects under the hood. | Replays the latest or a specified number of previous emissions to new subscribers.<br><br>It uses ReplaySubject under the hood. |
| **Unsubscription logic** | Unsubscribes when the last subscriber unsubscribes. | Offers a refCount option to unsubscribe when the last subscriber unsubscribes. By default, refCount is set to false. However, if you keep it set to false, the source Observable will remain active even when the subscriber count reaches zero. This situation can be risky because if the source Observable never completes, it might lead to memory leaks. |

Figure 10.3 – share and shareReplay comparison table

In RxJS 7, the share operator was enhanced with an optional configuration object as an argument, share(config), which makes it more flexible and ready to do the job of other operators, such as shareReplay(). In this configuration object, there are four properties:

- Connector: With this option, you can control whether or not share will replay emissions. You can choose the subject type you're connecting through (such as ReplaySubject).

- resetOnRefCountZero: With this option, you can control when your Observable should be reset. If this option is enabled and all the subscriptions of our Observable are unsubscribed, the Observable is reset. However, if this option is disabled, the subject will remain connected to the source.

- `resetOnComplete`: If enabled, the resulting Observable will reset on completion.

- `resetOnError`: If enabled, the resulting Observable will reset after an error.

So, `shareReplay` is nothing but a `share` operator that uses `ReplaySubject` as a connector and a specific reset strategy.

The following code shows how we can achieve the behavior of the optimized `shareReplay` operator by using the `share` operator instead:

```
recipes$ = timer$.pipe(
  switchMap(_ =>
    this.http.get<Recipe[]>(`${BASE_PATH}/recipes`)),
  share({
    connector: () => new ReplaySubject(),
    resetOnRefCountZero: true,
    resetOnComplete: true,
    resetOnError: true
  })

);
```

The preceding code shows the `share` operator with the same behavior as the `shareReplay` operator. This is because we referenced `ReplaySubject` as a connector, so we're telling `share` to use replay logic.

Then, for the reset strategy, we enabled all the reset options – `resetOnRefCountZero`, `resetOnComplete`, and `resetOnError` – to get optimized behavior and enhanced performance.

That's it – by using the `share` operator, we can achieve the same behavior as the `shareReplay` operator!

---

**Note**

Apart from the `shareReplay` operator, a lot of work was done in RxJS 7 to consolidate multicasting operators. The `multicast`, `publish`, `publishReplay`, `publishLast`, and `refCount` operators were deprecated and will be removed in RxJS 8, and the only operators that will remain are `shareReplay`, `share`, and `connectable`.

As we have seen in this section, the `share` operator rules them all, meaning that in most cases, it is highly recommended to use the `share` operator instead of `connectable` and `shareReplay`. The `shareReplay` operator is too popular to deprecate but may be deprecated in future versions as there is an alternative to it, especially because `shareReplay`, when not used carefully, can cause memory leaks, particularly with infinite streams.

So, if you're using RxJS 7, it is highly recommended to call the `share` operator instead of `shareReplay`.

Now that we've learned how we can optimize HTTP requests by caching our data using `shareReplay` and `share` operators and have put those operators in place in `RecipesApp` to cache the list of recipes, let's discover another situation where caching streams is very useful.

## Highlighting the use of caching for side effects

The use case we covered in this chapter involved optimizing HTTP requests to enhance our web applications' performance. All you have to do is put the result in a cache, which acts as a shared place for all consumers.

There are other use cases where caching streams makes a lot of sense, namely when accounting for expensive side effects on the streams. In general, we call the actions that we perform after a value is emitted **side effects**. This could be logging, displaying messages, doing a mapping, and so on.

Here's an example of a side effect using the `tap` operator:

```
import {map, from } from 'rxjs';
import { tap } from 'rxjs/operators';

const stream$ = from([1, 2, 'Hello', 5]);
stream$

  .pipe(
    tap((value) => console.log(value)),
    map((element) => {
      if (isNaN(element as number)) {
        throw new Error(element + ' is not a number');
      }
      return (element as number) * 2;
    })
  )
  .subscribe({
    next: (message) => console.log(message),
    error: (error) => console.log(error),
    complete: () => console.log('Stream Completed'),
  });

//console output
1
2
2
4
Hello
Error
```

In the preceding code, we are performing a transformation for every number that's emitted, multiplying it by 2, and returning the multiplied value. If the value is not a number, an error is thrown. However, we need to log the initial value before the transformation. That's why we called the `tap` operator before the `map` operator – so that we can log the original value. This is a basic example of a side effect, but others could also occur, such as handling errors or displaying messages.

> **Note**
>
> For further details about the `tap` operator, please refer to the official documentation: `https://rxjs.dev/api/operators/tap`.

In some situations, side effects can perform other actions that are more complex than logging, such as displaying messaging and handling errors. This can include some computations that represent an expensive treatment in terms of performance. Unfortunately, every subscriber will execute those treatments, even though it is enough to run them only once. Otherwise, it will harm the performance of your application.

If you have this use case in your application, it is highly recommended that you use the `share` operator to cache the result and execute heavy treatments only once.

## Summary

In this chapter, we explained various caching concepts in web applications, including their benefits and use cases. We focused on a concrete example in our recipe app, detailed the requirement, and implemented it reactively. Through this, we learned about the behavior of the `shareReplay` operator, as well as the alternative implementation – that is, using the `share` operator in RxJS 7. Finally, we highlighted how caching can help us when we have heavy side effects in our app.

In the next chapter, we will explore the reactive pattern for bulk operations.

# 11

# Performing Bulk Operations

**Bulk operations** are tasks performed on a large scale, such as uploading many files at once, deleting or inserting many items in one shot, or applying a transformation or computation to multiple elements of a list simultaneously.

These operations are designed to handle multiple updates in a single operation, often resulting in improved efficiency and performance compared to when each item is processed separately. Tracking the progress of bulk operations is crucial to provide feedback to users, monitor the health of the operation, and identify potential issues.

In this chapter, we will start by explaining the bulk operation requirement and the type of bulk operation that we will consider. After that, we will walk you through the different steps to implement the reactive pattern for implementing bulk operations. Finally, we will learn the reactive pattern for tracking the bulk operation's progress.

In this chapter, we're going to cover the following main topics:

- Defining the bulk operation requirements
- Learning the reactive pattern for bulk operations
- Learning the reactive pattern for tracking the bulk operation's progress

## Technical requirements

This chapter assumes that you have a basic understanding of RxJS.

The source code of this chapter is available at `https://github.com/PacktPublishing/Reactive-Patterns-with-RxJS-and-Angular-Signals-Second-Edition/tree/main/Chap11`.

## Defining the bulk operation requirements

In web applications, a bulk operation is represented by one action or event; however, in the background, there are two possible behaviors:

- Running one network request for all the tasks
- Running parallel network requests for every task

In this chapter, we will be using the second behavior. We want to allow the user to upload the recipe images at once, track the progress of the upload operation, and display a progress bar to the user. We can see what this will look like here:

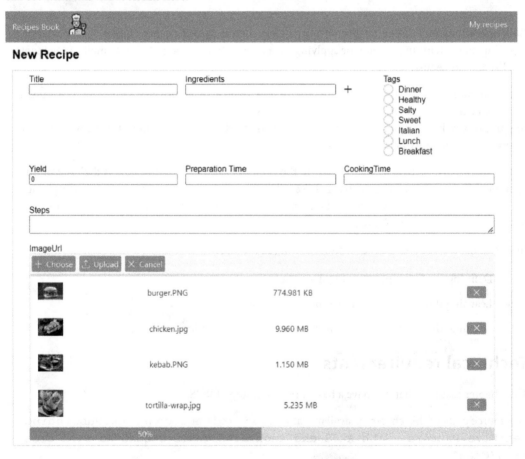

Figure 11.1 – Uploading the recipe's images

In the `RecipeCreation` interface, we will be changing the layout of the **ImageUrl** field to the **File Upload** layout available in our library of components, PrimeNG, as shown in the screenshot. The **File Upload** layout allows the user to choose multiple files, clear the selection, and upload the files.

The upload will be done on the server, and we have a specific service for the upload that takes both the file to be uploaded and the identifier of the associated recipe as input. Since the backend upload API supports only one file at a time, we will be running $N$ network requests in parallel to upload $N$ files (i.e., if we upload two files, two requests will be sent). This is the bulk change use case that we will consider in this chapter.

In the UI, we will have one event that will trigger multiple requests at the same time. The following diagram provides a graphical representation of the bulk operation:

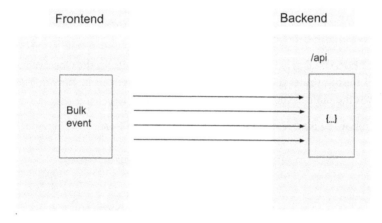

Figure 11.2 – A bulk operation visualization

So, to sum up, we want to do the following:

- Allow the user to upload many files after clicking only once on the **Upload** button
- Display the progress of this bulk operation

Now that we have defined the requirement, let's see how we can implement it in a reactive way.

## Learning the reactive pattern for bulk operations

As usual, we have to consider our tasks as streams. As the task that we are going to perform is uploading the recipe image in the backend, let's imagine a stream called `uploadRecipeImage$` that will take the file and the recipe identifier as input and perform an HTTP request. If we have $N$ files to be uploaded, then we will create $N$ streams.

We want to subscribe to all those streams together, but we are not interested in the values emitted from each stream through the process. Instead, we only care about the final result (the last emission) – whether the file is uploaded successfully, or something wrong happens and the upload fails.

Is there an RxJS operator that gathers a list of Observables together to get a cumulative result? Thankfully, yes: we have the `forkJoin` operator.

## The forkJoin operator

The `forkJoin` operator falls under the category of combination operators. If we look at the official documentation, we find this definition:

*"Accepts an Array of ObservableInput or a dictionary Object of ObservableInput and returns an Observable that emits either an array of values in the exact same order as the passed array, or a dictionary of values in the same shape as the passed dictionary."*

In other words, `forkJoin` takes a list of Observables as input, waits for the Observables to complete, and then combines the last values they emitted in one array and returns it. The order of the values in the resulting array is the same as the order of the input Observables.

Let's consider the following marble diagram to better understand this:

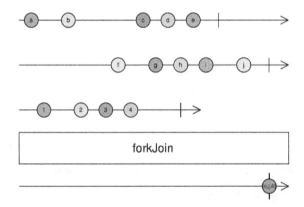

Figure 11.3 – A forkJoin marble diagram

Here, `forkJoin` has three input Observables (represented by the three timelines before the operator box).

The first Observable emitted the **a** value, but `forkJoin` does not emit anything (look at the last timeline after the operator box, which represented the returned result by `forkJoin`).

Then, the third Observable emitted **1**, and, again, nothing was emitted by `forkJoin`. Why? Because, as we said in the definition, `forkJoin` will emit only once when all the Observables are complete.

So, as illustrated in the marble diagram, `forkJoin` emitted only once when the last Observable (the second one) completed. Let's break this down:

- The third Observable (represented by the third timeline) completed first, and the last value emitted was **4**.

- Then, the first Observable (represented by the first timeline) completed, and the last value emitted was **e**. At that time, `forkJoin` did not emit any value because there was still an Observable running.

- Finally, the last Observable (represented by the second timeline) completed, and the last value emitted was **j**. Therefore, `forkJoin` returns an array containing the results of each stream in the order of the input Observables (**e**, **j**, and **4**).

The order of completion is not considered; otherwise, we would have had `[4,e,j]`. Even though the third Observable was completed before the first and second one, `forkJoin` respected the order of the input Observables and returned the **e** value before the **4** and **j** values.

So, keep in mind that `forkJoin` emits once when all the input Observables are complete and preserves the order of the input Observables.

This fits our requirements well! `forkJoin` is best used when you have a list of Observables and only care about the final emitted value of each. That's what we want to do. In our case, we will issue multiple upload requests, and we only want to take action when a response is received from all the input streams.

Let's now see the bulk operation reactive pattern in action.

## The bulk operation reactive pattern

To utilize the pattern in our recipe app, first, we need to create a new service called `UploadRecipesPreviewService` under `src/app/core/services`, which is responsible for uploading the files. Here is the service's code:

```
import { HttpClient } from '@angular/common/http';
import { Injectable } from '@angular/core';
import { Observable } from 'rxjs';
import { UploadStatus } from '../model/upload.status.model';
import { environment } from 'src/environments/environment';
const BASE_PATH = environment.basePath

@Injectable({
  providedIn: 'root'
})
export class UploadRecipesPreviewService {

  constructor(private http: HttpClient) { }
```

```
upload(recipeId: number|undefined|null, fileToUpload:
File): Observable<UploadStatus> {
  const formData = new FormData()
  formData.append('fileToUpload', fileToUpload as File)
  return this.http.post< UploadStatus >(
    `${BASE_PATH}/recipes/upload/${recipeId}`,
    formData
  )
}
}
```

The upload method issues the HTTP upload request and returns the upload status (whether having succeeded or failed). This method takes two parameters as input:

- recipeId: The identifier of the recipe
- fileToUpload: The file to be uploaded

Then we used FormData to send the file to the server. FormData is an object in JavaScript that allows you to easily build a set of key-value pairs representing form fields and their values respectively.

Now we need to implement the behavior of the **Upload** button. For this purpose, in the RecipeCreationComponent template, we need to specify the method that will be called when clicking on the **Upload** button – the onUpload method in our case – and put it as a value to the callback – uploadHandler – provided by the component library we are using to get triggered when the user uploads the files. Here's the HTML template snippet:

```
<div class="form-row">
    <div class="col-12">
        <label for="ImageUrl">ImageUrl</label>
        <p-fileUpload name="imageUrl" [multiple]=true
            [customUpload]="true" (uploadHandler)=
                "onUpload($event.files)">
        </p-fileUpload>
    </div>
</div>
```

> **Note**
>
> Some of the code in the template has been removed here for brevity. You can find the full template code in the book's GitHub repository, the link for which can be found in the *Technical requirements* section.

Next, we need to implement the `onUpload` method and define our reactive streams in `RecipeCreationComponent`. So, we will define the following:

- A `BehaviorSubject` that will always emit the last value of the uploaded files, called `uploadedFilesSubject$`, and initialize it with an empty array:

```
uploadedFilesSubject$ = new BehaviorSubject<File[]>([]);
```

- The `onUpload (files: File[])` method, which is called when clicking on the **Upload** button. Here, we will update the emissions of `uploadedFilesSubject$` with the last array of uploaded files as follows:

```
onUpload(files: File[]) {
  this.uploadedFilesSubject$.next(files);
}
```

- A stream called `uploadRecipeImages$` that is responsible for doing the bulk upload as follows:

```
uploadRecipeImages$ =
this.uploadedFilesSubject$.pipe(
    switchMap(uploadedFiles=>forkJoin(
    uploadedFiles.map((file: File) =>
      this.uploadService.upload(
      this.recipeForm.value.id, file))))
)
```

Let's break down what's going on in the code here, piece by piece.

Every time we click on the **Upload** button, `uploadedFilesSubject$` will emit the files to be uploaded. We need to listen to `uploadedFilesSubject$` emissions, and then use `switchMap` (which we learned about in *Chapter 6, Transforming Streams*) to transform every value emitted by `uploadedFilesSubject$` to the Observable that will be built using `forkJoin`.

To `forkJoin`, we pass an array of the Observables responsible for uploading each file. We built the array of Observables by mapping every file in the `uploadedFiles` array to the stream, resulting from calling the `upload` method available in `UploadRecipesPreviewService` that takes the `id` property of the recipe (which we retrieved from `recipeForm`) and the file as input.

Now that we've established our upload logic and defined the upload stream, it's time to subscribe to the `uploadRecipeImages$` stream. We need to inject `UploadRecipesPreviewService` into the constructor and subscribe to `uploadRecipeImages$` in the template, as follows:

```
<ng-container *ngIf="uploadRecipeImages$ | async"></ng-
  container>
```

Now, let's suppose one of the inner streams errors out. The `forkJoin` operator will no longer emit any values for us. This is another important thing to be aware of when using this operator. You will lose the value of any other stream that would have already been completed if you do not catch the error correctly on the inner Observable. Therefore, catching the error in this case is crucial!

This is how we handle it:

```
uploadRecipeImages$ = this.uploadedFilesSubject$.pipe(
  switchMap(uploadedFiles=>forkJoin(uploadedFiles.map((
    file: File) =>
    this.uploadService.upload(this.recipeForm.value.id,
      file).pipe(
        catchError(errors => of(errors)),
  )))))
```

Here, we called `catchError` on the inner stream returned by the `upload` method. Then, we wrapped the error inside another Observable and returned it. This way, the `forkJoin` stream will stay alive and emit values.

It makes a lot of sense to catch the errors in order to display something significant to the user – for example, in our case, if one of the uploads fails because the maximum image file size was reached or the extension of the image is not allowed, then the system should display such an exception to the user to help them fix the file.

## Benefits of the forkJoin operator

To sum up, `forkJoin` has the following benefits:

- It is very useful when you are interested in combining results and getting a value only once
- It only emits *once*, when all the Observables complete
- It preserves the order of the input Observables in the emission
- It will complete when one of the streams errors out, so make sure you handle the error

Now, at this point, our code works nicely. But what if we need to know some information during the process, such as how many files were already uploaded? What is the progress of the operation? How much time do we still need to wait?

With the current `forkJoin` implementation, it is not possible, but let's see how we can do it in the next section.

# Learning the reactive pattern for tracking the bulk operation's progress

Tracking the progress of bulk operations is very important, as it provides feedback to the user and can identify potential issues. When it comes to approaches for tracking progress, there are different strategies and techniques depending on the nature of the bulk operation and the technology stack you're using. For example, you can use an increment counter to show when each operation is processed, use a percentage to track the progress of the operations, or even log the progress to a file or database.

In the case of our recipe app, in order to track the progress of the bulk upload, we will use the percentage of completion strategy. To implement this strategy, we will use a very useful operator called `finalize`.

The `finalize` operator allows you to call a function when the Observable completes or errors out. The idea is to call this operator and execute a function that will calculate the progress. This way, every time an Observable completes, the progress will get updated.

This is what the code will look like:

```
counter: number = 0;
uploadProgress: number=0;

uploadRecipeImages$ = this.uploadedFilesSubject$.pipe(
  switchMap(uploadedFiles =>
  forkJoin(uploadedFiles.map((file: File) =>
    this.uploadService.upload(this.recipeForm.value.id,
      file).pipe(
        catchError(errors => of(errors)),
        finalize(() => this.calculateProgressPercentage(
          ++this.counter, uploadedFiles.length))
      ))))
)

  private calculateProgressPercentage(completedRequests:
  number, totalRequests: number) {
    this.uploadProgress =
      Math.round((completedRequests / totalRequests) *
        100);
  }
onUpload(files: File[]) {
  this.uploadProgress=0;
  this.counter=0;
  this.uploadedFilesSubject$.next(files);
}
```

The finalize operator calls the calculateProgressPercentage private function that takes the following parameters:

- *The number of completed requests*: We just declare a counter property that we will increment every time the Observable completes
- *The total number of requests*: This number is retrieved from the array of uploadedFiles

Inside the calculateProgressPercentage function, we perform a simple computation to identify the completion percentage and store the result in an uploadProgress property. When the user clicks on **Upload**, both the uploadProgress and counter properties should be reset to 0.

Then, you can map the value of this property to any ProgressBar component in the UI. In our case, we used the PrimeNG p-progressBar component as follows:

```
<div class="row">
    <div class="col-12">
        <label for="ImageUrl">ImageUrl</label>
        <!-- <input type="text" name="imageUrl"
        formControlName="imageUrl"> -->
        <p-fileUpload name="imageUrl" [multiple]=true
            [customUpload]="true"
            (uploadHandler)="onUpload($event.files)"
            accept="image/*"></p-fileUpload>
        @if(uploadProgress>0) {
        <p-progressBar [value]=uploadProgress>
            </p-progressBar>
        }
    </div>
</div>
```

Here, we only display p-progressBar when the upload is in progress (uploadProgress>0) and we pass the uploadProgress value as input to the progress component. This way, you will be able to display the progress to the user.

Here is the result in our app:

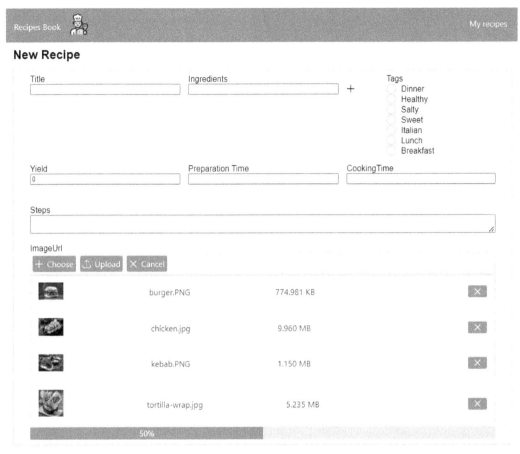

Figure 11.4 – The file upload progress bar

## Summary

In this chapter, we explained the concept of bulk operation and learned how to implement a real-world example of a bulk task in a reactive way. We learned the behavior and a use case of the `forkJoin` operator and went through the different steps to implement a bulk upload. Finally, we went through a reactive technique to implement the tracking progress functionality using the `finalize` operator.

In the next chapter, we will explore the pattern of real-time updates and the different techniques available in RxJS to implement them at the lowest cost.

# 12

# Processing Real-Time Updates

**Real time** refers to the capability of an application to handle and respond to data or events immediately as they happen, without any noticeable delay or latency. This is a very hot topic nowadays, with the demand for real-time features growing in web applications, particularly in areas such as live financial trading, live tracking systems, and live monitoring, analytics, and healthcare. Ultimately, the quicker you get the data, the sooner you can react and make decisions, increasing the chances of higher profits.

So, how can you process real-time messages in the frontend and update the displayed data automatically in the UI? This is what we will cover in this chapter. We will start by explaining the real-time requirement, and then we will walk you through the different steps to implement the reactive pattern for consuming real-time updates. Finally, we will learn the reactive pattern for handling reconnection.

In this chapter, we're going to cover the following main topics:

- Defining the requirements of real time
- Learning the reactive pattern for consuming real-time messages
- Learning the reactive pattern for handling reconnection

## Technical requirements

This chapter assumes that you have a basic understanding of RxJS.

We used the ws library, which is a WebSocket Node.js library, in order to support WS in our backend. For more details, check out this link: `https://github.com/websockets/ws`.

The source code of this chapter is available at `https://github.com/PacktPublishing/Reactive-Patterns-with-RxJS-for-Angular-16-2nd-Edition/tree/main/Chap12`.

# Defining the requirements of real time

There are two techniques available for publishing real-time data on the web:

- **Pull technique**: This is where the client raises a request to get the latest version of data. **HTTP polling** and **HTTP long polling** are two examples of implementations of this pull technique.

- **Push technique**: This is where the server pushes updates to the client. **WebSockets** and **server-sent events** are two implementations of this push technique.

We are not going to discuss or compare these techniques in detail, as it is not the goal of this chapter; however, in general, push techniques have a lower latency compared to pull ones. For this reason, we will use the push technique and WebSocket as the implementation for our requirements.

In short, the WebSocket protocol is a stateful communication protocol that establishes a low-latency bi-directional communication channel between a client and a server. In this way, messages can be sent back and forth between the server and the client.

The following diagram illustrates the WebSocket communication flow:

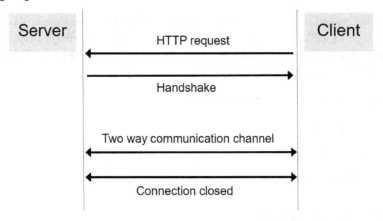

Figure 12.1 – WebSocket communication

As illustrated, there are three steps in WebSocket communication:

1. **Opening the connection**: In this step, the client issues an HTTP request to tell the server that a protocol upgrade will occur (from HTTP to WebSocket). If the server supports WebSockets, then the protocol switch will be accepted.

2. **Establishing the communication channel**: Once the protocol upgrade is done, then a bi-directional communication channel will be created, and messages start to be sent back and forth between the server and the client.

3. **Closing the connection**: When the communication is over, a request will be issued to close the connection.

At this level, that is all that you need to know about WebSockets. Now, let's quickly review what we'll be doing in our app.

In our recipe app, `RecipesListComponent` is responsible for displaying the list of recipes. We will simulate the addition of a new recipe (a recipe for chilli chicken) after a delay of 5 seconds following the rendering of `RecipesListComponent`. The UI should then be updated instantly to include this new recipe by rendering it in the `RecipesList` page.

You will find a ready-to-use WebSocket backend under the `recipes-book-api` folder; this is what pushes the new recipe to the frontend 5 seconds after establishing the connection. We will also use a timer in the backend to simulate the arrival of a new recipe. `RecipesListComponent` should then consume the message coming from the WebSocket server and push the newly received recipe in the already displayed list of recipes. The UI should be updated automatically without having to trigger any **Refresh** button to get the updates.

So, without further ado, in the next section, let's see how we can implement all of this using RxJS's `WebSocketSubject`.

# Learning the reactive pattern for consuming real-time messages

RxJS has a special type of subject called `WebSocketSubject`; this is nothing but a wrapper around the W3C WebSocket object, which is available in the browser. It allows you to communicate with a WebSocket server, both sending and consuming data through a WebSocket connection.

Let's explore the capabilities of `WebSocketSubject` and learn how to use it to consume real-time messages in our project.

## Creating and using WebSocketSubject

In order to use `WebSocketSubject`, you have to call the `webSocket` factory function that produces this special type of subject and takes the endpoint of your WebSocket server as input. The following is the function signature:

```
webSocket<T>(urlConfigOrSource: string | WebSocketSubjectConfig<T>):
WebSocketSubject<T>;
```

It accepts two types of arguments, either of the following:

- A string representing the URL of your WebSocket server endpoint
- A special object of the `WebSocketSubjectConfig` type that contains the URL of your endpoint, along with other properties (we will explore `WebSocketSubjectConfig` in detail in the *Learning the reactive pattern for handling reconnection* section)

The following code is an example of calling the `webSocket` factory function with the first type of argument:

```
import { webSocket } from "rxjs/webSocket";
const subject = webSocket("ws://localhost:8081");
```

The next piece of code is an example of calling the `webSocket` factory function using the second type of argument:

```
import { webSocket } from 'rxjs/webSocket';
const subject$ = webSocket({url:'ws://localhost:8081'});
```

In our case, the URL of our endpoint is `ws://localhost:8081`. You can use `wss` for a secure WebSocket connection (which is the same as HTTPS for a secure HTTP connection).

We will be using both types of arguments in this chapter.

Let's now see how we can establish the connection to the WebSocket in the following section.

### Opening the connection

Now that you have a reference for `WebSocketSubject`, you should subscribe to it:

```
import { webSocket } from 'rxjs/webSocket';
const subject$ = webSocket({url:'ws://localhost:8081'});
subject$.subscribe();
```

This will establish the connection with your ws endpoint and allow you to start receiving and sending data. Of course, if you don't subscribe, the connection will not be created.

### Listening to incoming messages from the server

`WebSocketSubject` is nothing but a regular RxJS subject, whereby you can register callbacks to listen and process the incoming messages from the WebSocket server.

In order to listen to messages, you should subscribe to the produced `WebSocketSubject` from the `webSocket` factory function and register a callback, as follows:

```
const subject$ = webSocket('ws://localhost:8080');

// Listen to messages from the server
const subscription = subject$.subscribe(msg => {
  console.log('Message received from the socket'+ msg);
});
```

Here, we're simply subscribing to the WebSocket subject to initiate a connection with the WebSocket server and then logging any received messages to the console.

### Pushing messages to the server

To send messages to the server, we just use the `next` method available in the `subject` type:

```
// Push messages to the server
subject$.next('Message to the server');
```

### Handling errors

You can also catch errors coming from the server using `catchError` as usual and push errors to the server by calling the `error` method. Here's an example:

```
// Push errors to the server
subject$.error('Something wrong happens')

// Handle incoming errors from the server
subject$.pipe(catchError(error=>of('Something wrong happens')))
```

However, bear in mind that when you send an error, the server will get notified about this error, and then the connection will be closed. So, nothing will get emitted thereafter.

### Closing the connection

You can use `unsubscribe` or `complete` to close the connection:

```
// Close the connection
subject$.complete();
//or
subject$.unsubscribe();
```

So, to wrap up what we've been discussing, only the creation of `WebSocketSubject` is specific to this special kind of subject. However, all the other APIs used (`subscribe`, `unsubscribe`, `complete`, `catchError`, `next`, and so on) are the same as those used for regular subjects. The following figure illustrates this whole process:

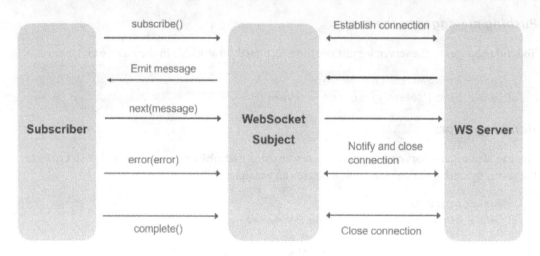

Figure 12.2 – WebSocketSubject possible events

Now that we've covered various WebSocket manipulations, from creating and establishing connections to sending messages, handling errors, and consuming incoming messages, let's explore a common pitfall you should keep in mind.

### Connection management

At this point, there is a particular behavior you should be aware of. If the same instance of WebSocketSubject has many subscribers, then they will share the same connection to save resources. However, if we have two different instances of WebSocketSubject, it will establish two distinct connections, even if they are referencing the same endpoint.

The following code explains the connection management for both use cases:

```
const firstSubject$ = webSocket('ws://localhost:8080');
const  secondSubject$ = webSocket('ws://localhost:8080');

// the first subscriber, opens the WebSocket connection
const subscription1 = firstSubject$.subscribe(msg => {
});
// the second subscriber, uses the already opened WebSocket
    connection
const subscription2 = firstSubject$.subscribe(msg => {
});
//this subscriber opens a new connection
const subscription3 = secondSubject$.subscribe(msg => {
});
```

Let's explain what's happening in this code. First, we create two instances of `WebSocketSubject` called `firstSubject$` and `secondSubject$`, respectively, which both reference the same ws endpoint.

Then, we create a subscription to `firstSubject$`; this first subscription will open the WebSocket connection. Then, we create a second subscription to the same Observable, `firstSubject$`; this second subscription will use the already opened WebSocket connection.

However, the subscription to `secondSubject$` will open a new WebSocket connection. Why? Because it is a new reference to the WebSocket subject, even though it references the same ws endpoint as `firstSubject$`.

Now, if we have many subscribers sharing the same connection and one of those subscribers decides to complete, then the connection will be released unless there are no more subscribers listening, as described in the following code block:

```
const subject$ = webSocket('ws://localhost:8080');
// the first subscriber, opens the WebSocket connection
const subscription1 = subject$.subscribe(msg => {});

// the second subscriber, uses the already opened WebSocket connection

const subscription2 = subject$.subscribe(msg => {});

// the connection stays open
subscription1.unsubscribe();

// closes the connection
subscription2.unsubscribe();
```

This is all that you need to know to make a basic scenario work. Simple, right?

Now, let's see the recommended pattern for putting our recipe app in place.

## WebSocketSubject in action

Now that we know how to create a connection to our ws endpoint, it is time to explore the different steps to consume real-time messages in our `RecipesApp`. In particular, we will establish a connection to the WebSocket server, and once the new recipe is sent to the frontend, we will update it in the UI. Let's delve into the various steps necessary to fulfill this requirement.

## Step one – create a real-time service

The first step is to isolate all the interactions with `WebSocketSubject` in a separate Angular service. To do this, we will create an Angular service called `RealTimeService` under the `src/app/core/services` path. `RealTimeService` will look like this:

```
import { Injectable } from '@angular/core';
import { webSocket, WebSocketSubject } from 'rxjs/webSocket';
import { environment } from '../../../environments/environment';
import { Recipe } from '../model/recipe.model';
export const WS_ENDPOINT = environment.wsEndpoint;
@Injectable({
  providedIn: 'root'
})
export class RealTimeService {
  private socket$: WebSocketSubject<Recipe[]> | undefined;
  private messagesSubject$ = new
    BehaviorSubject<Observable<Recipe[]>>(EMPTY);
  private getNewWebSocket(): WebSocketSubject<Recipe[]> {
    return webSocket(WS_ENDPOINT);
  }
  sendMessage(msg: Recipe[]) {
    this.socket$?.next(msg);
  }
  close() {
    this.socket$?.complete();
  } }
```

Let's break down what's happening at the level of this code in the service that we defined:

- We have a private property, `socket$`, of type `WebSocketSubject<Recipe[]>|undefined`, as we will receive an array containing one or more recipes from the backend. `socket$` contains the reference to the WebSocket subject that we will create using the `getNewWebSocket()` method.

- We have a private `BehaviorSubject` named `messagesSubject$`, which is responsible for transmitting the latest incoming messages from the WebSocket server to new subscribers. We've provided the type `Observable<Recipe[]>` for `messagesSubject$`, as it will emit an Observable containing an array of recipe objects. Initially, we've set it to `EMPTY`, which is an Observable that immediately completes without emitting any values.

- We have a private method, `getNewWebSocket()`, that calls the `webSocket` factory function, passing a constant named `WS_ENDPOINT` as input, and this returns `WebSocketSubject`.

  `WS_ENDPOINT` represents the endpoint of the WebSocket server defined in the `src/environments/environment.ts` file as `wsEndpoint`. Note that URLs are environment-specific configurations, which means they can change from one environment to another (e.g., development, staging, and production). Defining endpoint URLs in the `environment.ts` file is a common practice in Angular applications because it provides a centralized location to handle environment-specific configuration settings, so you can easily switch between environments without modifying your application code.

- We have a public method, `sendMessage()`, that sends a message that is given as input to the socket, which will forward the message to the server.

- Finally, we have a public method, `close()`, that closes the connection by completing the subject.

Then, we will add the `connect()` method, which will listen to the incoming messages in a reactive way and emit messages to the subscribers as follows:

```
public connect(): void {
    if (!this.socket$ || this.socket$.closed) {
    this.socket$ = this.getNewWebSocket();
    const messages = this.socket$.pipe(
      tap({
        error: error => console.log(error),
      }), catchError(_ => EMPTY));
    this.messagesSubject$.next(messages);
  }
}
```

Let's break down what is going on in this method. If `socket$` is undefined (not yet created) or closed, then `socket$` will be populated by the newly produced `WebSocketSubject` from the `getNewWebSocket` method.

Then, we will combine the `tap` and `catchError` operators; the `tap` operator is used to log a message when an error occurs or when the connection closes, and the `catchError` operator handles errors and returns an empty Observable.

The returned Observable from the pipe operation will be stored in a constant called `messages`. The `messagesSubject$` Observable will emit the messages' Observable (so, it is an Observable of Observables):

After that, we will provide a read-only copy from the `messagesSubject$` Observable through the `messages$` public Observable defined in `RealTimeService` as follows: :

```
public messages$ = this.messagesSubject$.pipe(
switchAll(), catchError(e => { throw e }));
```

We used the `SwitchAll` operator to flatten the Observable of an Observable, and we will be subscribing to `messages$` in every component that needs to consume real-time updates. Why do we do this? The idea is to protect `Subject$` and the incoming messages from any external update and expose the messages to the consumers as read only. In this way, any component interested in consuming the real-time messages has to subscribe to `messages$`, and all the logic related to the socket will be handled privately in this service.

### Step two – trigger the connection

After putting the service in place, we should call the `connect` method. As we want the `connect` method to be triggered just once, we will call it from the root component, `src/app/app.component.ts`, after injecting `RealTimeService`. Here's the code to be added:

```
constructor(private service: RealTimeService ) {
this.service.connect();
}
```

### Step three – define the Observable emitting live updates

Next, we should call the `messages$` Observable in the adequate Angular component. As we want to update the list of recipes with the newest ones, we should define the Observable in `RecipesListComponent`.

But wait! We already have an Observable in `RecipesListComponent` named `recipes$` that fetches the list of recipes from `RecipesService`:

```
recipes$=this.service.recipes$;
```

Could we use this existing Observable instead of creating a new one? Absolutely!

Our goal is to initially display the list of recipes emitted by `recipes$` and then seamlessly incorporate any newly added recipes emitted by `messages$`. We can achieve this using the `combineLatest` operator from RxJS.

The `combineLatest` operator merges the latest values from multiple Observables into an array and emits a new array whenever any of the source Observables emits a value. By leveraging this operator, we can combine `recipes$` and `messages$` as follows:

```
recipes$=combineLatest([this.service.recipes$,
this.realTimeservice.messages$]).pipe(map(([recipes,
```

```
updatedRecipes]) => {
  // Merge or concatenate the two arrays into a single
     array
  return [...recipes, ...updatedRecipes];
}));
```

In the code, we combined `recipes$` and `messages$` and then used the `map` operator to extract the latest values emitted by each. We then merge these values into a single array, which is then returned. This ensures that `recipes$` consistently emits a unified array containing all recipes.

### Preventing data loss with the scan operator

Now, let's quickly consider a scenario where a recipe with an ID of 12 is initially pushed and added to the recipes list. If another recipe with, for example, an ID of 14 is pushed afterward from the server, then the newest pushed recipe (ID 14) will override the previous one (ID 12). Therefore, the ID 12 recipe will be lost. To prevent this data loss, we can use the `scan` operator.

The `scan` operator in RxJS is similar to the reduce function in JavaScript. It applies an accumulator function over an Observable sequence and returns each intermediate result, emitting the accumulated value each time a new value is emitted by the source Observable. In simpler terms, it continuously applies a function to each value emitted by the source Observable, accumulating these values over time and emitting the intermediate results. This operator is useful for maintaining the state, accumulating values, or performing any kind of stateful transformation on Observable streams.

So, in our case, we can use the `scan` operator as follows:

```
recipes$ = combineLatest([
  this.service.recipes$,
  this.realTimeService.messages$
]).pipe(
  scan((acc: Recipe[], [recipes, updatedRecipes]:
  [Recipe[], Recipe[]]) => {
    // Merge or concatenate the two arrays into a single
       array
    return acc.length === 0 &&
      updatedRecipes.length === 0 ? recipes : [...acc,
        ...updatedRecipes,];
  }, [])
);
```

In this context, `scan` ensures that all emitted recipes, including both the initial recipes fetched from the `this.service.recipes$` stream and any subsequent updates received from `this.realTimeService.messages$`, are accumulated into a single array. This prevents the loss of data that could occur if a simple mapping operation were used. As a result, the `recipes$` Observable stream contains a comprehensive and up-to-date list of recipes, reflecting all changes from both sources throughout its lifetime.

### Step four – subscribe to the Observable emitting live updates

Finally, we just have to subscribe to the `recipes$` Observable in our component's template using the async pipe, which is already carried out in `recipes-list.component.html`:

```
@if ( recipes$ | async; as recipes) {
  ....
}
```

However, we have one more tweak to consider! Since we've established that `messages$` emits after a 5-second delay following the emission of `recipes$`, there's a slight problem: `combineLatest` only emits once both Observables have emitted values.

To circumvent this brief latency while waiting for `messages$` to emit, in `RealTimeService`, we can use the `startWith()` operator on the `messages$` subject to supply an initial value of an empty array as follows:

```
public messages$ =
this.messagesSubject$.pipe(switchAll(), startWith([]),
catchError(e => { throw e }));
```

After executing this code, you will notice that 5 seconds after displaying 11 recipes, the recipe with an ID of `12` (chili chicken) will be added to the list on the second page of our cards list. If another recipe is pushed afterward, it will be accumulated to the current list of recipes.

Note that in the case of frequent updates in the UI, it is highly recommended to set the change detection strategy to `onPush` in order to optimize the performance, like so:

```
@Component({
  selector: 'app-recipes-list',
  standalone: true,
  changeDetection: ChangeDetectionStrategy.OnPush
})
```

And that's it! You will be able to consume live updates in a reactive way using this pattern.

At this point, you may be wondering how to handle reconnection. When the server is restarted or the connection crashes for whatever reason, does this subject restore the lost connection under the hood?

The answer is **no**. The reconnection capability is not provided by `WebSocketSubject`.

However, you can implement this easily in your web application using RxJS. Let's learn how you can do this in the next section.

## Learning the reactive pattern for handling reconnection

When the connection to the WebSocket server is lost, the channel will be closed, and `WebSocketSubjet` will no longer emit values. This is not the expected behavior in the world of real time. The reconnection capability is a must in most cases.

Therefore, let's imagine, for example, that after a disconnection, a system tries to reconnect after every 3 seconds. The solution, in this case, is intercepting the closure of the socket and retrying the connection. How can we intercept the closure of the connection?

This is possible thanks to `WebSocketSubjectConfig`, which is responsible for customizing some behavior in the socket life cycle. The `WebSocketSubjectConfig` interface in RxJS provides several properties that you can use to configure a WebSocketSubject. These properties allow you to customize various aspects of WebSocket communication:

```
export interface WebSocketSubjectConfig<T> {
  url: string;
  protocol?: string | Array<string>;
  /** @deprecated Will be removed in v8. Use {@link
  deserializer} instead. */
  resultSelector?: (e: MessageEvent) => T;
  openObserver?: NextObserver<Event>;
  serializer?: (value: T) => WebSocketMessage;
  deserializer?: (e: MessageEvent) => T;
  closeObserver?: NextObserver<CloseEvent>;
  closingObserver?: NextObserver<void>;
  WebSocketCtor?: { new(url: string,
  protocols?:string|string[]): WebSocket };
  binaryType?: 'blob' | 'arraybuffer';
}
```

Let's explain the different properties available in `WebSocketSubjectConfig`:

- `url`: This property specifies the URL of the WebSocket endpoint to connect to (we've already explained and and used this in this chapter).

- `protocol`: This property specifies the subprotocol to use during the WebSocket handshake (refer to *Figure 12.1*). It can be a single string or an array of strings representing the subprotocols.

- `resultSelector`: This property specifies a function that takes the WebSocket event as input and returns the value to be emitted by `WebSocketSubject`. It's commonly used to extract specific data from WebSocket events; however, it is deprecated and will be removed in version 8 of RxJS.

- `closeObserver`: This property specifies an observer object that listens for the WebSocket connection closing. It can be used to handle cleanup tasks or perform actions when the connection is closed.

- `openObserver`: This property specifies an observer object that listens for the WebSocket connection opening. It can be used to perform actions when the connection is successfully established.

- `binaryType`: This property specifies the binary type of WebSocket messages. It can be either of the JavaScript types `blob` or `arraybuffer`. By default, it's set to `blob`.

- `serializer`: This property specifies a function used to serialize outgoing messages before sending them over the WebSocket connection. It's commonly used to convert objects or complex data structures into strings.

- `deserializer`: This property specifies a function used to deserialize incoming messages received over the WebSocket connection. It's commonly used to parse received strings back into objects or other data types.

These properties provide flexibility and control over WebSocket communication in RxJS. You can customize them according to your specific requirements to optimize WebSocket interactions in your application.

> **Note**
>
> The full description of each property is available in the official documentation link: `http://bit.ly/RxJS-WebSocket`.

In order to benefit from `WebSocketSubjectConfig`, you should call the `webSocket` factory function, which takes the second type of parameter. The following code creates `WebSocketSubject` using `WebSocketSubjectConfig` and simply intercepts the closure event to display a custom message:

```
private getNewWebSocket() {
  return webSocket({
    url: WS_ENDPOINT,
    closeObserver: {
      next: () => {
        console.log('[RealTimeService]: connection
                     closed');
      }
    },
```

```
    });
  }
```

Now that we know how to intercept the closure of the connection, let's learn how to retry the reconnection. We can combine the `retryWhen` operator that conditionally resubscribes to an Observable after it completes using the `delayWhen` operator that sets the delay between two consecutive connections.

So, let's create a function that will retry to connect to a given Observable for every configurable RECONNECT_INTERVAL; we will log into the browser's console on every attempt at reconnection:

```
private reconnect(observable: Observable< Recipe[] >):
Observable< Recipe[] > {
  return observable.pipe(retryWhen(errors =>
    errors.pipe(
      tap(val => console.log('[Data Service]
        Try to reconnect', val)),
          delayWhen(_ => timer(RECONNECT_INTERVAL)))));
}
```

This `reconnect` function will be used as an RxJS custom operator to handle the reconnection after the socket's closure in the `connect()` method of our `RealTimeService`, as follows:

```
public connect(cfg: { reconnect: boolean } = { reconnect: false }):
void {

  if (!this.socket$ || this.socket$.closed) {
    this.socket$ = this.getNewWebSocket();
    const messages = this.socket$.pipe(cfg.reconnect ?
    this.reconnect : o => o,
      tap({
        error: error => console.log(error),
      }), catchError(_ => EMPTY))
    this.messagesSubject$.next(messages);
  }
}
```

As you can see, a new Boolean `reconnect` parameter is added to the `connect` function to differentiate between the reconnection and the first connection. This optimizes the code and avoids adding an additional function.

Then, all you have to do is call the `connect` function with `reconnect: true` when intercepting the connection closure:

```
private getNewWebSocket() {
  return webSocket({
    url: WS_ENDPOINT,
```

```
        closeObserver: {
          next: () => {
            console.log('[DataService]: connection
                        closed');
            this.socket$ = undefined;
            this.connect({ reconnect: true });
          }
        },
      });
```

In this way, after the connection closure, you will see many outgoing requests from the client trying to reach the server every 3 seconds.

The reconnection capability is a must in the world of real time. This is how we handled it using RxJS in a few lines of code. Many developers don't know that RxJS offers this feature, which enables you to consume real-time messages coming from WebSocket and add many third-party libraries to handle this requirement, and it is also available out of the box. So, choosing RxJS, in this case, is one less dependency!

## Summary

In this chapter, we delved into a practical demonstration of consuming real-time messages from a WebSocket server in a reactive manner. We first outlined the requirements and provided context for the implementation. Subsequently, we explored the capabilities of WebSocketSubject and described the step-by-step process, from establishing a connection to handling incoming messages from the socket.Next. We applied these concepts to a real-world scenario within the recipe app, gaining insights into best practices for implementing real-time functionality and ensuring robust connection control.

Finally, we expanded our understanding by incorporating a reconnection mechanism in a reactive way, leveraging the WebSocketSubjectConfig and RxJS operators to achieve seamless connection management.

Now, as we approach the final chapter of this book, let's switch gears and focus on testing Observables.

# Part 5:
# Final Touches

In this part, you'll discover the different strategies to test reactive streams. We'll explore their benefits and when to use each one, reinforcing your learning with practical examples.

This part includes the following chapter:

- *Chapter 13, Testing RxJS Observables*

13

# Testing RxJS Observables

Observables play a central role in managing asynchronous data streams and event-driven interactions. By thoroughly testing Observables, developers can verify the correctness of their asynchronous code, anticipate and handle various edge cases, and ensure consistent behavior across different environments and use cases.

The comprehensive testing of Observables not only enhances the robustness of applications but also improves code quality, reduces the likelihood of bugs and regressions, and ultimately enhances the overall user experience. With rigorous testing practices in place, developers can confidently deploy reactive applications that meet high standards of reliability, performance, and usability.

Many developers consider testing Observables a challenging task. This is true. However, if you learn the right techniques, you can implement maintainable and readable tests in a very effective manner.

In this chapter, we will walk you through three commonly used patterns for testing streams. We will start by explaining the subscribe and assert pattern, after which we will discuss the marble testing pattern. Finally, we will highlight a suitable pattern for testing streams that are returned from `HTTPClient` by focusing on a concrete example in our recipe app.

In this chapter, we're going to cover the following main topics:

- Learning about the subscribe and assert pattern
- Learning about the marble testing pattern
- Highlighting testing streams using `HTTPClientTestingModule`

## Technical requirements

This chapter assumes that you have a basic understanding of RxJS and unit testing in Angular using Jasmine. Follow this link for more information: `https://angular.dev/guide/testing#set-up-testing`.

> **Note**
>
> angular.dev will be the new documentation site for Angular developers; it offers updated features and documentation. angular.io will be deprecated in future releases.

We will be testing Observables in an Angular context. The source code for this chapter is available at https://github.com/PacktPublishing/Reactive-Patterns-with-RxJS-for-Angular-16-2nd-Edition/tree/main/Chap13.

We will be completing a unit test for the saveRecipe method, which is available under the RecipesService class. You can find the complete code in the recipes.service.spec file.

## Learning about the subscribe and assert pattern

As you will already know, Observables are lazy, and we don't obtain any value until we subscribe to them. In tests, it is the same thing; Observables will not emit any value until we subscribe to them. To solve this, programmers always tend to subscribe to the Observables manually inside the tests and then perform assertions on the emitted values. This is what we call the **subscribe and assert** pattern.

Let's delve into testing using the subscribe and assert pattern across three distinct scenarios. We will demonstrate testing for methods returning a single value, methods returning multiple values, and methods returning timed values (values returned after a specified time duration).

### Testing single-value output methods

Let's suppose we have to test a method that returns a single value. The method is called getValue(value: boolean) and is available in an Angular service called SampleService.

The method itself is very simple, returning an Observable that will emit the Boolean input value as follows:

```
import { Observable, of } from 'rxjs';
export class SampleService {

  getValue(value: boolean): Observable<boolean> {
    return of(value);
  }

}
```

The test of this method will look like this:

```
describe('SampleService', () => {
  let service: SampleService;
  beforeEach(() => {
```

```
  service = TestBed.inject(SampleService);
});

it('should return true as a value', () => {
  service.getValue(true).subscribe(
    result=>expect(result).toEqual(true))
});
});
```

Here, we start by defining our test suite using the Jasmine `describe()` function. The function is used to define a test suite, which is a logical grouping of test cases to execute a single task with different test scenarios. It serves as a way to organize and structure your tests, making them more readable and maintainable.

The `describe()` function takes two parameters:

- A string description of the test suite (in the previous code snippet, `SampleService`, which is the name of the service that we are going to test and refers to the description of our test suite).

- A function that contains the test cases for that suite (in testing frameworks, "test case" and "spec" typically refer to a single unit of testing within a test suite). Inside this function, we inject `SampleService` into the `beforeEach` statement to provide a shared instance of the service that we will be using in all the test cases. Finally, we define the test case of our `getValue(value: boolean)` method by using the Jasmine function `it()`. The `it()` function takes two parameters:

  - A string description of the test case (in the previous code snippet, `should return true as a value` refers to the description of our test case).

  - A function that contains the test logic. In this function, we subscribe to the `getValue(true)` method and expect the result to be equal to `true` since we passed the `true` value as input. Expectations are built using the `expect()` Jasmine function and are used to assert or verify that certain conditions are met during the execution of a test.

Now, let's run `ng test`; the test passes, and everything is fine:

```
TOTAL: 2 SUCCESS
✓ Browser application bundle generation complete.
Chrome 95.0.4638.69 (Windows 10): Executed 2 of 8 (skipped 6) SUCCESS (0.046 secs / 0.018 secs)
TOTAL: 2 SUCCESS
✓ Browser application bundle generation complete.
Chrome 95.0.4638.69 (Windows 10): Executed 2 of 8 (skipped 6) SUCCESS (0.046 secs / 0.016 secs)
TOTAL: 2 SUCCESS
✓ Browser application bundle generation complete.
Chrome 95.0.4638.69 (Windows 10): Executed 1 of 7 (skipped 6) SUCCESS (0.038 secs / 0.018 secs)
TOTAL: 1 SUCCESS
```

Figure 13.1 – ng test output

Quite simple, right? This is the expected behavior when running a positive scenario. A positive scenario typically involves providing input or conditions that align with the expected behavior of the code being tested, resulting in successful execution without errors or failures.

Now, let's handle a negative scenario by providing input or conditions that are intended to trigger failures in the code being tested. To do so, we will replace `true` with `false` in the following assertion:

```
it('should return true as a value', () => {
service.getValue(true).subscribe(
  result=>expect(result).toEqual(false))
});
```

When you run `ng test` again, our test will fail.

However, in some cases, the test will still pass. How is this possible?

The thing with expectations in testing is that if you have an unmet assertion, it throws an error. Plus, if you have an unhandled error inside an RxJS subscription, it will be thrown on a separate call stack, meaning that it is asynchronously thrown. Therefore, tests that use the subscribe and assert pattern can sometimes be green even though, in reality, they are failing.

To overcome this, we should pass a `done` callback to the test function and call it manually after our expectations for when the test has been completed, as follows:

```
it('should return true as a value', (done) => {
  service.getValue(true).subscribe(
    result => {
      expect(result).toEqual(false);
      done();
    }
  );
});
```

The `done` callback is a mechanism used in asynchronous testing to signal the completion of a test case to the testing framework. It is supported by many testing frameworks, such as Jasmine, Jest, and Mocha. Calling the `done` callback ensures that the test doesn't finish prematurely before all asynchronous tasks have been executed and assertions have been verified. Therefore, we prevent false positives and ensure that our tests accurately reflect the behavior of the code being tested, particularly in asynchronous scenarios. So, don't forget to call the `done` callback in asynchronous scenarios!

Now, let's consider a more complicated method that will return multiple values instead of one value.

## Testing multiple-value output methods

Let's consider a method called `getValues`, which will return multiple values like so:

```
export class SampleService {

  getValues(): Observable<String> {
    return of('Hello', 'Packt', 'Readers');
  }
}
```

The values will be emitted one by one in the aforementioned order.

When using the assert and subscribe pattern, the test will look like this:

```
it('should return values in the right order', (done) => {
  const expectedValues = ['Hello', 'Packt', 'Readers'];
  let index = 0;
  service.getValues().subscribe(result => {
    expect(result).toBe(expectedValues[index]);
    index++;
    if (index === expectedValues.length) {
      done();
    }
  });
});
```

In the preceding code, we created an array that represents the expected values in order; then, we subscribed to the `getValues` method and compared the emitted values with the expected value using a counter (`expectedValues[index]`). After finishing, we called the `done()` callback.

However, instead of the counter, we can use the `toArray` operator of RxJS, which will put the values that have been emitted in an array and then compare the resulting array with the expected array we defined:

```
it('should return values in the right order', (done) => {
  const expectedValues = ['Hello', 'Packt', 'Readers'];
  service.getValues().pipe(toArray()).subscribe(result => {
    expect(result).toEqual(expectedValues);
    done();
  });
});
```

Well, this is working fine, and ng test will pass. However, in both cases, even though we are dealing with a simple stream, we were forced to add some logic; in the first example, we added a counter, while in the second example, we used the toArray operator. This boils down the tests and adds some unnecessary testing logic; these are the most significant drawbacks of the subscribe and assert pattern.

Now, let's move on to a different example and explore testing methods that output timed values.

## Testing timed-value output methods

Let's update the method getValues() and add a timer to return the values after a specific duration as follows:

```
getValues(): Observable<String> {
  return timer(0, 5000).pipe(
    take(3),
    switchMap((result) => of('Hello', 'Packt',
                             'Readers'))
  )
```

Here, we used the RxJS timer in this method to emit a value every 5 seconds. Since timer produces an endless stream, we call the take operator to return the first three emissions and complete them. Then, for every emission, we use the switchMap operator to return an Observable that emits three values consecutively.

This is tricky, right? If we use the subscribe and assert pattern here, the tests would be very complicated and may take a lot of time, depending on the value that's passed to timer. However, the unit tests should be fast and reliable.

In this case, having a virtual timer can be beneficial. A virtual timer refers to a simulated passage of time controlled by the testing framework. Instead of waiting for actual time to pass, which could lead to slow and unreliable tests, the virtual timer allows testers to control time programmatically. This means they can advance time forward or backward as needed to trigger certain events or test scenarios, making it easier to write reliable and deterministic tests for code that depends on time-based behavior. This approach ensures that tests are fast, predictable, and independent of real-time conditions.

So, in a nutshell, the subscribe and assert pattern is a valid and easy technique that most developers adopt. However, it has some drawbacks that I pointed out throughout this section:

- We need to remember to call the done callback in asynchronous tests; otherwise, the tests will return invalid results.

- In some scenarios, we end up with boiled tests and unwanted testing logic.

- Timed Observables are very complicated to test.

Now, let's explore another approach for testing Observables: marble testing with RxJS testing utilities.

# Learning about the marble testing pattern

Marble diagrams are very useful for visualizing Observable execution. You will already know this, as we introduced marble diagrams back in *Chapter 1*, *Diving into the Reactive Paradigm*, and we've used them in almost all the reactive patterns we've implemented in this book. They are simple to understand and delightful to read. So, why not also use them in code? What you will be surprised to know is that RxJS introduced marble testing as an intuitive and clean way to test Observables.

Let's discover what marble testing is about. We will start by explaining the syntax in the next section and then learn how we can write marble tests in our code.

## Understanding the syntax

To understand the syntax, we should know about the following semantics:

| Character | Meaning |
| --- | --- |
| ' ' | This represents a special character that will not be interpreted. It can be used to align your marble string. |
| '-' | This represents a frame of virtual time passing. |
| '\|' | This represents the completion of an Observable. |
| [a-z] | This represents a value that is emitted by an Observable. It is an alphanumeric character. |
| '#' | This represents an error. |
| '()' | This represents a group of events that occur in the same frame. It can be used to group any values emitted, errors, and completion. |
| '^' | This represents the subscription point and will only be used when you're dealing with hot Observables. |
| [0-9] + [ms\|s\|m] | This represents time progression and allows you to progress virtual time by a specific amount. It's a number, followed by a time unit in milliseconds (ms), seconds (s), or minutes (m) without any space between them. |

Figure 13.2 – Marble testing syntax

This is the basic syntax. Let's look at some examples to practice the syntax:

- ---: This represents an Observable that never emits.
- -x--y--z|: This represents an Observable that emits x on the first frame, y on the fourth, and z on the seventh. After emitting z, the Observable completes.

- `--xy--#`: This represents an Observable that emits `x` on frame two, `y` on frame three, and an error on frame six.

- `-x^(yz)--|`: This is a hot Observable that emits `x` before the subscription.

You've got the idea, right? Now, let's learn how to implement marble tests in our code.

## Introducing TestScheduler

There are different packages out there that can help you write marble tests, including `jasmine-marbles`, `jest-marbles`, and `rxjs-marbles`. However, RxJS provides testing utilities out of the box, and all the libraries are just wrappers around the RxJS testing utilities. I recommend working with the RxJS utilities for the following reasons:

- You don't have to include a third-party dependency

- You stay up to date with the core implementation

- You stay up to date with the latest features

The RxJS API that's provided for testing is based on `TestScheduler`. This API allows you to test time-dependent RxJS code in a controlled and deterministic manner, which is crucial for writing reliable and predictable tests for Observables with time-based operators.

To define our test logic, the `TestScheduler` API provides a `run` method that has the following signature:

```
run<T>(callback: (helpers: RunHelpers) => T): T;
```

The `run` method takes a `callback` function as an argument. This `callback` function is where you define your test logic, including setting up Observables, defining expectations, and making assertions. The `callback` function takes one argument named `helpers` of type `RunHelpers`, which provides various utility functions and properties to assist you in writing marble tests for Observables.

The `RunHelpers` interface contains the following properties:

```
export interface RunHelpers {
cold: typeof TestScheduler.prototype.
createColdObservable;
hot: typeof TestScheduler.prototype.
createHotObservable;
flush: typeof TestScheduler.prototype.flush;
expectObservable: typeof TestScheduler.
prototype.expectObservable;
expectSubscriptions: typeof TestScheduler.
prototype.expectSubscriptions;
}
```

Let's look at these properties one by one:

- `cold`: This produces a cold Observable based on a given marble diagram. Here is the signature of the method:

```
/**
    * @param marbles A diagram in the marble DSL.
      Letters map to keys in `values` if provided.
    * @param values Values to use for the letters in
      `marbles`. If ommitted, the letters themselves
      are used.
    * @param error The error to use for the `#`
      marble (if present).
    */
createColdObservable<T = string>(marbles: string,
values?: {
        [marble: string]: T;
    }, error?: any): ColdObservable<T>;
```

- `hot`: This produces a hot Observable based on a given marble diagram. Here is the signature of the method:

```
/**
    * @param marbles A diagram in the marble DSL.
      Letters map to keys in `values` if provided.
    * @param values Values to use for the letters in
      `marbles`. If ommitted, the letters themselves
      are used.
    * @param error The error to use for the `#`
      marble (if present).
    */
createHotObservable<T = string>(marbles: string,
values?: {
        [marble: string]: T;
    }, error?: any): HotObservable<T>;
```

When you're creating a hot Observable, you can use ^ to point out the first frame:

- `flush`: This starts virtual time. It's only needed if you use helpers outside the `run` callback or if you want to use `flush` more than once.

- `expectObservable`: This asserts that an Observable matches a marble diagram.

- `expectSubscriptions`: This asserts that an Observable matches the expected subscriptions.

Now, let's learn how we can implement marble testing using `TestScheduler` in the following section.

## Implementing marble tests

In this section, we will consider implementing marble tests for the `getValues` method previously mentioned in the subscribe and assert pattern:

```
export class SampleService {

  getValues(): Observable<String> {
    return of('Hello', 'Packt', 'Readers');
  }
}
```

The steps for writing the marble testing implementation pattern are simple:

1.  Import `TestScheduler` from `rxjs/testing`:

    ```
    import { TestScheduler } from 'rxjs/testing';
    ```

2.  In the `beforeEach` statement, inject `SampleService`. Then, instantiate `TestScheduler` and pass an input function that compares the actual output with the expected output of the Observable:

    ```
    import { TestScheduler } from 'rxjs/testing';
    describe('Service: SampleService', () => {
      let scheduler : TestScheduler;
      let service: SampleService;



      beforeEach(() => {

          service = TestBed.inject(SampleService);

          scheduler = new TestScheduler((actual, expected) => {

          expect(actual).toEqual(expected);

        });
      });
    });
    ```

    If the expected output and actual output are not equal, it throws an error, failing the test.

3.  Use `TestScheduler` to test your stream by calling the `run` method and passing a callback to it (remember that the callback needs to accept `RunHelpers` as the first parameter):

```
it('should return values in the right order', () => {
  scheduler.run((helpers) => {
  });
});
```

It is also useful to destruct the helpers into variables and use them directly to implement the marble tests. We will be destructuring the `expectObservable` variable, as we will use it to assert that the Observable matches the marble diagram, as follows:

```
it('should return values in the right order', () => {
  scheduler.run(({expectObservable}) => {
  });
});
```

4.  Finally, declare the expected marble and values and perform the expectation:

```
it('should return values in the right order', () => {
  scheduler.run(({expectObservable}) => {
    const expectedMarble = '(abc|)' ;
    const expectedValues = {a:'Hello', b:'Packt',
                            c:'Readers'};
    expectObservable(service.getValues()).toBe(
      expectedMarble, expectedValues)
  });
});
```

The `expectedMarble` constant represents the marble diagram. Since the `getValues` method returns three values consecutively, we used parentheses to group the a, b, and c emissions. The stream then completes, so we use the | character.

The `expectedValues` constant represents the values of the a, b, and c characters that we put in `expectedMarble`. It represents `'Hello'`, `'Packt'`, and `'Readers'`, consecutively, which are nothing but the values that are emitted by the Observable that we want to test.

The last instruction is the expectation; we should provide the expected result that our methods should return. Here, we must use `expectObservable`, which takes the Observable we want to test as a parameter and matches it with `expectedMarble` and `expectedValues`.

That's it. Let's have a look at the complete test setup:

```
describe('SampleService marble tests', () => {
  let scheduler : TestScheduler ;
  let service: SampleService;

  beforeEach(() => {
```

```
    service = TestBed.inject(SampleService);
    scheduler = new TestScheduler((actual, expected) => {
    expect(actual).toEqual(expected);
});
});

it('should return values in the right order', () => {
    scheduler.run(({expectObservable}) => {
    const expectedMarble = '(abc|)' ;
    const expectedValues = {a:'Hello', b:'Packt',
                            c:'Readers'};
    expectObservable(service.getValues()).toBe(
        expectedMarble, expectedValues)
    });
});
});
```

When you run `ng test`, this test will pass. If you enter wrong values in `expectedValues`, the test will fail:

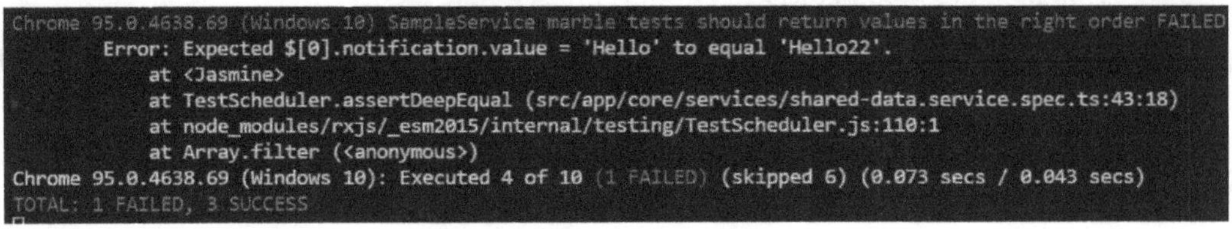

```
Chrome 95.0.4638.69 (Windows 10) SampleService marble tests should return values in the right order FAILED
        Error: Expected $[0].notification.value = 'Hello' to equal 'Hello22'.
            at <Jasmine>
            at TestScheduler.assertDeepEqual (src/app/core/services/shared-data.service.spec.ts:43:18)
            at node_modules/rxjs/_esm2015/internal/testing/TestScheduler.js:110:1
            at Array.filter (<anonymous>)
Chrome 95.0.4638.69 (Windows 10): Executed 4 of 10 (1 FAILED) (skipped 6) (0.073 secs / 0.043 secs)
TOTAL: 1 FAILED, 3 SUCCESS
```

Figure 13.3 – ng test failing

Well, this is cleaner than the subscribe and assert pattern implementation.

Now, let's look at a more difficult example and see how we can implement it using marble testing.

## Testing timed-value output methods

We will consider the testing of a timed Observable that was complicated to implement using the subscribe and assert pattern. Let's revisit the timer example that we explained earlier in the subscribe and assert pattern section:

```
getValues(): Observable<String> {
    return timer(0, 5000).pipe(
        take(3),
        switchMap((result) => of('Hello', 'Packt',
        'Readers'))
```

```
    )
  }
```

The cool `TestScheduler` feature that can help us here is **virtual time**; this allows us to test asynchronous streams synchronously by virtualizing time and ensuring that the correct items are emitted at the correct time. Thanks to the time progression syntax, we can advance virtual time by milliseconds (ms), seconds (s), or even minutes (m). This is extremely useful in the case of timed Observables.

Let's consider the following marble diagram:

```
e 999ms (fg) 996ms h 999ms (i|)';
```

Here, the diagram indicates that `e` is emitted immediately. Then, after 1 second, `f` and `g` are emitted. Then, 1 second later, `h` is emitted, after which `I` is emitted, and the stream finally completes.

Why use `999` and `996`? Well, we're using `999` because `e` takes 1 ms to emit and `996` because the characters in the `(fg)` group take 1 ms each.

With all this in mind, the marble tests of `getValues` will look like this:

```
const expectedMarble ='(abc) 4995ms (abc) 4995ms
   (abc|)' ;
```

The group of values `(abc)` is emitted every 5 seconds or 5000 ms, and since the characters are counted inside the group, we put `4995ms`. So, the whole test case will look like this:

```
it('should return values in the right time', () => {
  scheduler.run(({expectObservable}) => {
  const expectedMarble ='(abc) 4995ms (abc) 4995ms (abc|)';
  const expectedValues = {a:'Hello', b:'Packt',
                          c:'Readers'};
  expectObservable(service.getValues()).toBe(
    expectedMarble, expectedValues)
  });
});
```

That's how we resolved the test of a timed Observable using marble tests.

Marble testing is extremely powerful and helpful. It allows you to test a very high level of detail and complicated things such as concurrency and timed Observables. It also makes your tests cleaner. However, it requires you to learn a new syntax, and it is not recommended for testing business logic. Marble testing was designed for testing operators with arbitrary time.

> **Note**
>
> For more details about marble testing, you can check out the official docs at `https://rxjs.dev/guide/testing/marble-testing`.

Now, let's highlight a very common pattern for testing business logic.

# Highlighting testing streams using HttpClientTestingModule

Observables that are returned from the HTTP client are frequently used in our Angular code, but how can we test those streams? Let's look at the pattern we can use to test those Observables. We will be shifting our focus away from general testing practices and narrowing our attention specifically to testing our recipe app.

Consider the following method inside `RecipeService`:

```
saveRecipe(formValue: Recipe): Observable<Recipe> {
  return this.http.post<Recipe>(
    `${BASE_PATH}/recipes`, formValue);
}
```

The `saveRecipe` method issues an HTTP request and returns an Observable of recipe. In order to test the output Observable, there is a very useful API that can be used: `HttpClientTestingModule`. This API allows us to test HTTP methods that use the HTTP client. It also allows us to easily mock HTTP requests by providing the `HttpTestingController` service. In short, it enables us to mock requests instead of making real API requests to our API backend when testing.

Let's see the steps required to test the `saveRecipe` method using the `HttpClientTestingModule`:

1. Before you can use `HttpClientTestingModule`, import and inject it in your `TestBed` in the `beforeEach` statement, as follows:

```
import { TestBed } from '@angular/core/testing';
import { HttpClientTestingModule} from '@angular/common/http/
testing';
describe('RecipesService', () => {
  beforeEach(() => {
    TestBed.configureTestingModule({
      imports: [HttpClientTestingModule],
    });
  });
});
```

2. Then, import and inject `HttpTestingController` and `RecipesService` and provide a shared instance of each to use in our tests:

```
import { TestBed } from '@angular/core/testing';
import { HttpClientTestingModule, HttpTestingController } from
'@angular/common/http/testing';
```

```
import { RecipesService } from './recipes.service';

describe('RecipesService', () => {

  let service: RecipesService;
  let httpTestingController: HttpTestingController;

  beforeEach(() => {
    TestBed.configureTestingModule({
      imports: [HttpClientTestingModule],
      providers: [RecipesService]
    });
    httpTestingController =
      TestBed.inject(HttpTestingController)
    service = TestBed.inject(RecipesService)
  });
});
```

3. Next, implement the test case of saving the recipe. We'll mock `saveRecipe` as follows:

```
it('should save recipe from API', () => {
  const recipeToSave : Recipe= {
    "id": 9,
    "title": "Lemon cake",
    "prepTime": 10,
    "cookingTime": 35,
    "rating": 3,
    "imageUrl": "lemon-cake.jpg"

  }
  const subscription =
  service.saveRecipe(recipeToSave)
    .subscribe(_recipe => {
      expect(recipeToSave).toEqual(_recipe, 'should
      check mock data')
    });
  const req = httpTestingController.expectOne(
    `/api/recipes`);
  req.flush(recipeToSave);
  subscription.unsubscribe();
});
```

Here, we created a constant called `recipeToSave`, which represents a mocked recipe that we will post to the server to be saved. Then, we subscribed to the `saveRecipe` method and passed `recipeToSave` to it as a parameter. Inside the subscription, we defined our expectations. Then, we called the `expectOne` method, which expects a single request that's been made to match a given URL (in our case, `/api/recipes`) and returns mock data using the `flush` method, which resolves the request by returning a mocked body. Finally, we released the subscription.

4.  The last step is to add an `afterEach()` block, in which we run the `verify` method of our controller:

```
afterEach(() => {
  httpTestingController.verify();
});
```

The `verify()` method ensures that there are no outstanding HTTP requests that have not been handled or flushed. When you make HTTP requests in your tests using `HttpClientTestingModule`, they are intercepted by `httpTestingController` instead of being sent over the network. The `verify()` method ensures that all requests have been properly handled and allows your tests to pass only if there are no pending requests remaining.

In summary, the `afterEach()` block with `httpTestingController.verify()` is used in Angular tests to clean up and verify that there are no unhandled HTTP requests left over after each test case. This helps ensure that your tests are isolated and reliable, without unexpected network interactions.

And that's it; the pattern for testing methods that issue HTTP requests is complete. Just run the `ng test` command and ensure everything works fine.

> **Note**
> `HttpClientTestingModule` is very useful in this use case. For more details, please refer to https://angular.dev/guide/testing/services#httpclienttestingmodule.

## Summary

In this chapter, I've elucidated three common approaches for testing Observables in RxJS and Angular. Each solution has its strengths and weaknesses, and there is no one-size-fits-all answer.

First, we learned about the subscribe and assert pattern, as well as its advantages and drawbacks. This pattern is straightforward to understand but may not cover all edge cases, especially when dealing with complex asynchronous behavior.

Then, we learned about the marble testing pattern and its syntax, features, advantages, and drawbacks. We studied a basic example and an example that uses virtual time to test timed Observables. Marble testing provides a visual representation of Observable behavior; it is suitable for testing complex asynchronous scenarios. However, it requires special syntax, meaning it may have a steep learning curve for beginners.

Finally, we learned about a pattern that we can use to test streams that are returned from the HTTP client. This pattern provides control over responses and doesn't rely on external APIs. However, it can be tedious to set up and maintain and may not accurately simulate real-world network behavior in some cases.

In conclusion, each testing approach offers its advantages and trade-offs. Depending on your project requirements, you can choose the solution that aligns best with your testing needs and project constraints.

At this point, our journey into reactive patterns is coming to an end. In this book, I tried to highlight the most used reactive patterns that solve a lot of recurrent use cases in web applications. You can use them immediately in your current projects, adapt them to your needs, or get inspired to create your own reactive pattern.

This book is not just about patterns, though; it is also about the reactive approach and how to switch your mindset from imperative to reactive thinking; in most chapters, this is why I've highlighted the classic pattern before the reactive one to provide you with a smooth transition between the two.

And with that, we reach the end of our journey together. Thank you for reading and embarking on this reactive adventure with me!

# Index

# W

www.packtpub.com

Subscribe to our online digital library for full access to over 7,000 books and videos, as well as industry leading tools to help you plan your personal development and advance your career. For more information, please visit our website.

## Why subscribe?

- Spend less time learning and more time coding with practical eBooks and Videos from over 4,000 industry professionals

- Improve your learning with Skill Plans built especially for you

- Get a free eBook or video every month

- Fully searchable for easy access to vital information

- Copy and paste, print, and bookmark content

Did you know that Packt offers eBook versions of every book published, with PDF and ePub files available? You can upgrade to the eBook version at packtpub.com and as a print book customer, you are entitled to a discount on the eBook copy. Get in touch with us at customercare@packtpub.com for more details.

At www.packtpub.com, you can also read a collection of free technical articles, sign up for a range of free newsletters, and receive exclusive discounts and offers on Packt books and eBooks.

# Other Books You May Enjoy

If you enjoyed this book, you may be interested in these other books by Packt:

**Spring Boot and Angular**

Devlin Basilan Duldulao, Seiji Ralph Villafranca

ISBN: 978-1-80324-321-4

- Explore how to architect Angular for enterprise-level app development
- Create a Spring Boot project using Spring Initializr
- Build RESTful APIs for enterprise-level app development
- Understand how using Redis for caching can improve your application s performance
- Discover CORS and how to add CORS policy in the Spring Boot application for better security
- Write tests to maintain a healthy Java Spring Boot application
- Implement testing and modern deployments of frontend and backend applications

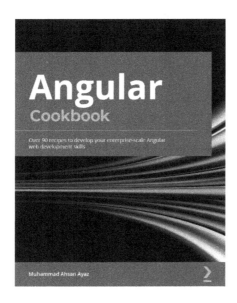

**Angular Cookbook**

Muhammad Ahsan Ayaz

ISBN: 978-1-83898-943-9

- Gain a better understanding of how components, services, and directives work in Angular
- Understand how to create Progressive Web Apps using Angular from scratch
- Build rich animations and add them to your Angular apps
- Manage your app's data reactivity using RxJS
- Implement state management for your Angular apps with NgRx
- Optimize the performance of your new and existing web apps
- Write fail-safe unit tests and end-to-end tests for your web apps using Jest and Cypress
- Get familiar with Angular CDK components for designing effective Angular components

## Packt is searching for authors like you

If you're interested in becoming an author for Packt, please visit `authors.packtpub.com` and apply today. We have worked with thousands of developers and tech professionals, just like you, to help them share their insight with the global tech community. You can make a general application, apply for a specific hot topic that we are recruiting an author for, or submit your own idea.

## Share Your Thoughts

Now you've finished *Reactive Patterns with RxJS and Angular Signals*, we'd love to hear your thoughts! Scan the QR code below to go straight to the Amazon review page for this book and share your feedback or leave a review on the site that you purchased it from.

https://packt.link/r/1-835-08770-1

Your review is important to us and the tech community and will help us make sure we're delivering excellent quality content.

# Download a free PDF copy of this book

Thanks for purchasing this book!

Do you like to read on the go but are unable to carry your print books everywhere?

Is your eBook purchase not compatible with the device of your choice?

Don't worry, now with every Packt book you get a DRM-free PDF version of that book at no cost.

Read anywhere, any place, on any device. Search, copy, and paste code from your favorite technical books directly into your application.

The perks don't stop there, you can get exclusive access to discounts, newsletters, and great free content in your inbox daily

Follow these simple steps to get the benefits:

1.  Scan the QR code or visit the link below

https://download.packt.com/free-ebook/9781835087701

2.  Submit your proof of purchase
3.  That's it! We'll send your free PDF and other benefits to your email directly

www.ingramcontent.com/pod-product-compliance
Lightning Source LLC
Chambersburg PA
CBHW080636060326
40690CB00021B/4955